SAFETY IN DESIGN

SAFETY IN DESIGN

C.M. van 't Land

Van 't Land Processing
Cort van der Lindenlaan 8
7521 AS Enschede
The Netherlands

Registered Office
John Wiley & Sons, Inc., 111 River Street, Hoboken, NJ 07030, USA

Editorial Office
111 River Street, Hoboken, NJ 07030, USA

For details of our global editorial offices, customer services, and more information about Wiley products visit us at www.wiley.com.

Library of Congress Cataloging-in-Publication Data
Names: Land, C.M. van 't, 1937- author.
Title: Safety in design / C.M. van 't Land, Van 't Land Processing.
Description: First edition. | Hoboken, NJ : John Wiley & Sons, Inc., [2018] |
   Includes bibliographical references and index. | Description based on
   print version record and CIP data provided by publisher; resource not
   viewed.
Identifiers: LCCN 2018016234 (print) | LCCN 2018019826 (ebook) | ISBN
   9781118745694 (Adobe PDF) | ISBN 9781118745588 (ePub) | ISBN 9781118745557
   (hardcover)
Subjects: LCSH: Nuclear reactors–Safety measures. | Nuclear reactors–Design
   and construction. | Chemical engineering–Safety measures.
Classification: LCC TK9152 (ebook) | LCC TK9152 .L36 2018 (print) | DDC
   621.48/35–dc23
LC record available at https://lccn.loc.gov/2018016234

Cover design by Wiley
Cover image: Courtesy of Connexxion Holding NV, Hilversum, The Netherlands

Set in 10.5/12.5pts TimesLTStd by SPi Global, Chennai, India

Printed in the United States of America

VWEP40592_111518

CONTENTS

PREFACE

This book emanates from the production of organic peroxides. The Dutch multinational Akzo Nobel, for which I worked as a chemical engineer between 1968 and 2000, manufactures these compounds.

In 1969, a Dutch company named Noury & Van der Lande became part of Akzo Nobel. That company had discovered around 1920 that dibenzoyl peroxide, a solid particulate material, can remove the yellowish color of flour. The finding was patented worldwide, licenses were given, and the industrial production of dibenzoyl peroxide was started. The production of synthetic plastics has increased since the 1940s, resulting in the increasing importance of organic peroxides as initiators of the radical polymerization of vinyl monomers. Noury & Van der Lande also started the production of organic peroxides for this application.

The expression "peroxides" is short for "superoxides." It indicates that the compound contains relatively much oxygen. All or part of this oxygen is "active oxygen". The active oxygen causes the desired action at the application of the organic peroxides. For example, the bleaching of flour is caused by the liberation of "active oxygen," oxidizing carotene to colorless compounds. A further example, at the manufacture of polymers, is the decomposition of organic peroxides at relatively low temperatures to form free radicals. The free radicals act as initiators for polymerization reactions.

Explosions and fires at the manufacture and the handling of these compounds have happened in the past. Peroxides are characterized by the presence of the peroxo group $-O-O-$. In organic peroxides, this group is bound to at least one carbon atom, or is bound to a carbon atom via a different atom. The presence of the peroxo group causes the thermal instability of organic peroxides. It also, in many instances, causes the sensitivity to impact, friction, and other chemicals. For example, dry dibenzoyl peroxide is very sensitive to impact, and serious accidents caused by this sensitivity have happened with this material in the past.

In retrospect, the most serious accidents within Noury & Van der Lande and Akzo Nobel occurred between 1935 and 1975. In this period, the production increased from tens to hundreds of metric tpa per product. The majority of serious accidents occurred during the reactions to produce organic peroxides.

My former colleague, the late Hans Gerritsen, proposed a method to improve the protection of the manufacture and handling of organic peroxides significantly. The method is called intrinsic continuous process safeguarding. The safeguarding is based on chemical and physical properties of reaction systems, and an activation of protection systems is not required. The method is also applicable to other chemical production systems. It is discussed in Chapter 1.

Hans Gerritsen also, at Deventer in The Netherlands in 1985, drew my attention to the fact that the methodology can be applied to all types of human activity, and that is what this book is about.

ACKNOWLEDGMENTS

I am grateful to Jan de Groot, who read the manuscript and, in doing so, made useful suggestions. Jan is the retired Head of Akzo Nobel's Safety Laboratory.

I am also grateful to retired professor Ad Verkooijen, who read Chapter 10 titled "Nuclear Power Stations". His comments enabled me to improve its contents.

Thanks are also due to many people providing information and figures. Their help was invaluable. Most people are open and supportive.

I am greatly indebted to my wife, Annechien, for her constant encouragement and patience.

C.M. VAN 'T LAND

1

INTRODUCTION

1.1 INTRODUCTION

A concept developed for the chemical industry can also be applied to other fields. This concept is called intrinsic continuous process safeguarding and is discussed in Section 1.2. It is related to the concept of inherently safer design. How the application of the concepts of inherently safer design and intrinsic continuous process safeguarding could have prevented three serious accidents in the chemical industry or mitigated its effects is briefly indicated in Sections 1.3–1.5. Section 1.6 contains concluding remarks.

1.2 INTRINSIC CONTINUOUS PROCESS SAFEGUARDING

The danger of explosions, evolution of toxic gases, etc., comes with the large-scale manufacture of certain chemicals. The prevention or control of undesirable reactions in processes is discussed in a paper [1]. The aim of intrinsic continuous process safeguarding is to obtain stable reaction systems that, within very wide limits, are not endangered by human errors or equipment failures. The approach has shown its merits at the manufacture of organic peroxides. It is related to the concept of inherently safer design [2]. Intrinsic continuous process safeguarding is compared to extrinsic process safeguarding in the paper mentioned earlier [1]. The latter safeguarding starts working upon a signal. Extrinsic process safeguarding is appropriate only as complementary and secondary protection: As complementary safeguarding by providing protection in places through which entering the hazardous area is improbable and as secondary protection by drawing up a second line of defense behind the intrinsic protection line.

Several serious accidents occurred in plants of the chemical industry in the second half of the previous century. Explosions, fires, and the emission of toxic materials were experienced. Three of these accidents will be discussed shortly in the following paragraphs. Kletz formulated the concept of inherently safer design, which encompasses hazard elimination and hazard reduction, for the first time in 1978 [3]. It was concerned with the safeguarding of the manufacture of chemicals. Our paper [1] also concerned the safeguarding of the manufacture of chemicals. The principles of these two related approaches can be used to formulate a generally applicable design strategy for the chemical industry. It is briefly indicated how the concepts of inherently safer design and intrinsic continuous process safeguarding could have either prevented the accidents in the chemical industry, described in the following paragraphs, or could have mitigated its effects.

1.3 THE FLIXBOROUGH ACCIDENT IN THE UNITED KINGDOM IN 1974

This accident occurred near a small village called Flixborough in a plant having a capacity of 70 000 tons of caprolactam per annum [4]. Caprolactam is an intermediate for the manufacture of Nylon 6 and Nylon 66. The village is in Lincolnshire and located south of Hull at England's east coast. The date of the accident is June 1, 1974. The accident comprised an explosion in the plant followed by fires. The name of the company involved was Nypro. It was jointly owned 55% by Dutch State Mines (DSM) and 45% by the National Coal Board (NCB) of England. Of those working on the site at the time, 28 were killed and 36 suffered injuries. Injuries and damages outside the works were widespread, but no one was killed. Fifty-three people were recorded outside the works as casualties. The 24-ha plant was almost completely destroyed. Outside the works, property damage extended over a wide area. The Report of the Court of Inquiry [4] states that the cause of the disaster was the ignition and rapid acceleration of deflagration, possibly to the point of detonation, of a massive vapor cloud formed by the escape of cyclohexane from the air oxidation plant under at least a pressure of $8.8 \, kg \, cm^{-2}$ and at a temperature of $155 \, °C$. In this plant, cyclohexane was, by means of a continuous process, converted into a mixture of cyclohexanol and cyclohexanone. Cyclohexanone was the intermediate product of the air oxidation plant. The Court estimates that the explosion was of the equivalent force to that of some 15–45 tons TNT. The cyclohexane oxidation plant contained six continuously stirred tank reactors in series. Prior to the accident, a reactor had to be removed for repair and the gap was bridged by a temporary 20-in. pipe, connected by a bellows at each end and inadequately supported on temporary scaffolding. The pipe collapsed. The escaping cyclohexane was a flashing liquid. At atmospheric pressure, its boiling point is $80.8 \, °C$. Approximately one-quarter of the escaping cyclohexane, having a temperature of $155 \, °C$, evaporated on escaping. The remaining three quarters thereby cooled down to, in principle, the boiling point at atmospheric pressure, that is, $80.8 \, °C$. Much of the remaining liquid formed a spray. The large cloud formed made the explosion possible. The source of the ignition was probably a hot surface in the hydrogen plant of the caprolactam plant.

Before 1972, cyclohexanone was produced at Flixborough via the liquid-phase hydrogenation of phenol. The latter process is a safer process than the air oxidation process. The reason is that it proceeds at temperatures below the atmospheric boiling point of the reaction liquids. Specifically, the boiling points at atmospheric pressure of phenol, cyclohexanol, and cyclohexanone are, respectively, 181.75, 161.1, and $156.5 \, °C$. From a safety point of view, the oxidation process introduced a new dimension. Large quantities of cyclohexane had to be circulated through the reactors under a working pressure of $8.8 \, kg \, cm^{-2}$ and at a temperature of $155 \, °C$. Any escape from the plant was therefore potentially dangerous. As stated above, the temporary 20-in. pipe in the oxidation plant was inadequately supported. However, a similar error in a liquid-phase phenol hydrogenation plant would not have had comparable consequences.

1.4 THE SEVESO EMISSION IN ITALY IN 1976

This accident occurred near a small village called Meda near Seveso, a town of about 17 000 inhabitants some 15 miles from Milan in Italy [5]. The accident happened on July 10, 1976. It comprised the emission of a white cloud drifting from the works from which materials settled out downwind. Among the substances deposited was a very small amount of

2,3,7,8-tetrachlorodibenzo-*p*-dioxin (TCCD), which is also known as dioxin, although there are more dioxins. This specific dioxin is one of the most toxic substances known. The process that gave rise to the accident was the production of 2,4,5-trichlorophenol (TCP) in a batch reactor. TCP is used for herbicides and antiseptics. The name of the company involved was ICMESA. It used a process developed by Givaudan, which was itself owned by Hoffmann La Roche. These last two companies are Swiss companies, whereas the former one is Italian.

People fell ill and animals died in the contaminated area over the days following July 7, 1976. People were evacuated from the area affected. There were no deaths of humans directly attributable to TCCD.

The reactor from which the emission took place was a 13 875-l vessel equipped with a stirrer and with a steam jacket supplied with steam at 12 bara. The boiling point of water at 12 bara is 188 °C.

The reactions to produce TCP had been started at 16.00 h on July 9, 1976. This date was a Friday. At 05.00 h on July 10, 1976, the batch was interrupted. The background was the closure of the plant for the weekend. At that point in time, the first chemical reaction had been completed. A distillation step followed the first chemical reaction; it comprised the removal of part of ethylene glycol (a solvent) from the reactor. The latter step had been started but had not been completed. The heat required for this distillation was supplied via a jacket. Steam entering the jacket came from a turbine. Because of the approaching weekend, the steam turbine was on reduced load and, although the steam pressure was 8 bara, its temperature had risen to about 300 °C. The interruption of the batch comprised the stopping of the heating and the stirring. At 05.00 h on July 10, 1976, the batch temperature was 158 °C. The upper section of the reactor wall, not wetted by the reactor contents, had, at that time, a temperature higher than 158 °C. The latter temperature was caused by the relatively high steam temperature. Based on this fact, Theofanous [6] proposed a sequence for the reaction runaway. The residual heat in the upper reactor section raised the temperature of the top layer of the liquid to 200–220 °C by radiation, a temperature high enough to initiate a runaway reaction leading to decomposition. Such a hot spot could develop because the stirring had been stopped. At 12.37 h on July 10, 1976, the bursting disk on the reactor ruptured and the emission took place.

The high temperature of the heating medium is, safetywise, an aspect. Noticeable decomposition reactions of the reaction mixture concerned already start at 185 °C. Limiting the temperature of the heating medium to, e.g. 165 °C, would have been appropriate. As to the manufacturing of TCP, it would have been better to bring the batch to completion. However, with a reduced heating medium temperature, the process would probably not have been endangered by human error.

1.5 THE BHOPAL EMISSION IN INDIA IN 1984

This accident occurred at Bhopal in India in a plant manufacturing carbamate pesticides [7, 8]. It is by far the worst accident that has ever occurred in the chemical industry. Bhopal is located in Central India in the state of Madhya Pradesh. At the time of the emission, the town had 800000 inhabitants. The plant was located at the outskirts of Bhopal. The date of the accident is December 3, 1984. The name of the company concerned was Union Carbide India Ltd (UCIL). The emission comprised the release of gaseous methyl isocyanate (MIC) through a nonfunctioning vent gas scrubber having a height of 30 m onto housing adjoining the site. The chemical is extremely toxic. MIC was an intermediate at the manufacture of Sevin, an insecticide. MIC could escape because it became inadvertently or deliberately contaminated with water in a storage tank. An exothermic reaction between MIC and water occurred. The reaction heat caused the evaporation of the compound. An aspect is that MIC's boiling point at atmospheric pressure is 38 °C. The rising pressure in the storage tank caused a relief valve to open. The inadvertent contamination with water due to a flushing (washing) operation is generally considered more probable than the deliberate contamination.

The number of people killed is officially 3787 [8] but is in actual fact much higher. Many more were wounded.

For the purpose of our present discussion, it is relevant to remark that a hazard and operability study of the plant might have revealed ways in which MIC could be contaminated by water. It would then be possible to prevent water to come into contact with MIC. Further main points are that a Sevin process route exists at which MIC is not obtained as an intermediate, that the intermediate storage was rather large, that several plant systems were not in working order, that the plant was not maintained properly, and that housing was too close to the plant.

1.6 CONCLUDING REMARKS

Intrinsic continuous process safeguarding is a safeguarding originating from the core of the process and is consequently directly and completely based on the reaction system and the reaction conditions; the safeguarding is based on chemical and physical properties [1].

Over time, people have invented and developed intrinsically protected approaches in many types of human activities. Two examples of such approaches will be discussed briefly. The first example concerns collecting mushrooms. The *Amanita phalloides* (a very toxic mushroom) may be mistaken for the champignon mushroom (edible). The color of both mushrooms tends toward white. An intrinsically protected, or, in other words, an inherently safer way of collecting mushrooms is to collect chanterelles, edible yellow mushrooms. The false chanterelles exist; however, they are edible, just not tasty. The Jack O'Lantern mushroom also appears similar to the chanterelle. The latter poisonous mushroom is usually found in woodland in North America. Although not lethal, consuming the Jack O'Lantern mushroom leads to strong complaints. Still, the collection of chanterelles is safer than the collection of champignon mushrooms.

The second example is given by Mannan [3]. A double-track railroad, with a dedicated track for each direction of travel, is inherently safer than a single track for both directions of travel.

REFERENCES

[1] Gerritsen, H.G. and van 't Land, C.M. (1985). Intrinsic continuous process safeguarding. *Industrial & Engineering Chemistry Process Design and Development* 24: 893–896.

[2] Mannan, S. (2005). *Lees' Loss Prevention in the Process Industries: Hazards Identification, Assessment, and Control*, 32/1–32/24. Amsterdam, Boston: Elsevier Butterworth-Heinemann.

[3] Mannan, S. (2005). *Lees' Loss Prevention in the Process Industries: Hazards Identification, Assessment, and Control*, 32/2–32/3. Amsterdam, Boston: Elsevier Butterworth-Heinemann.

[4] Court of Inquiry (1975). *The Flixborough Disaster*. London: Her Majesty's Stationary Office.

[5] Mannan, S. (2005). *Lees' Loss Prevention in the Process Industries: Hazards Identification, Assessment, and Control*, Appendix 3/1–3/13. Amsterdam, Boston: Elsevier Butterworth-Heinemann.

[6] Theofanous, T.G. (1983). The physicochemical origins of the Seveso accident – I. *Chemical Engineering Science* 38: 1615–1629.

[7] Mannan, S. (2005). *Lees' Loss Prevention in the Process Industries: Hazards Identification, Assessment, and Control*, Appendix 5/1–5/11. Amsterdam, Boston: Elsevier Butterworth-Heinemann.

[8] Pietersen, C.M. (2009). *After 25 Years: The Two Largest Industrial Disasters Concerning Dangerous Substances, LPG Disaster Mexico-City and Bhopal Tragedy*, 63–91. Nieuwerkerk aan den IJssel, The Netherlands: Gelling Publishing (in Dutch).

2

PROCEDURAL, ACTIVE, AND PASSIVE SAFETY

2.1 INTRODUCTION

How the safety in the chemical industry can be improved by the application of intrinsic continuous process safeguarding was discussed in Chapter 1. The concept was compared with extrinsic process safeguarding, which starts working upon a signal. It is, for other fields in society, useful to distinguish between procedural, active, and passive safety. Their definitions are given in Section 2.2. In Section 2.3, four examples of emergency power units that failed to come into action are dealt with. Three examples concern hospitals and one example a chemical plant. An emergency power unit is an active safety measure, as it starts working upon a signal. The failure of the blowout preventer (BOP) (an active safety measure) during the Gulf Oil accident in 2010 is discussed in Section 2.4. Section 2.5 deals with the safeguarding of Formula One races by means of mainly passive safety measures. Finally, Section 2.6 discusses explosion panels, also called bursting disks. These parts are designed to give in, if, due to a dust explosion, the subsequent pressure in a piece of equipment surpasses a predetermined value. Safeguarding by these components is continuously present.

Safety in Design, First Edition. C.M. van 't Land.
© 2018 John Wiley & Sons, Inc. Published 2018 by John Wiley & Sons, Inc.

2.2 DEFINITIONS

The definitions in this paragraph are borrowed from Kletz' and Amyotte's book [1]. A procedural safety method is a method activated by a human. The extinction of a fire by a fireman is an example. Of course, to avoid fires, preventive measures should be considered first. The use of materials that cannot take fire is an example. Complete cities burned down in the middle ages because the houses were made out of wood. Still, we cannot completely avoid the occurrence of fires, and to cope with the effects by means of a procedure is a possibility. However, the fire brigade may come in too late.

An active safety method is activated by a signal. For instance, in case of a fire, a water spray is turned on automatically by a smoke, flame, or heat detector. However, the equipment may fail or be turned off.

Both procedural safety methods and active safety methods can be compared to the concept of extrinsic process safeguarding used in the chemical industry as described in Chapter 1.

Finally, a passive safety method is immediately available. In case of a fire, fire-proof insulation is continuously available and does not need activation by humans or equipment. Passive safety methods can be compared to intrinsic continuous process safeguarding as described in Chapter 1.

Generally speaking, passive safety measures are better than active safety measures because they do not need activation. Active safety measures are better than procedural safety measures because they are already present.

2.3 FOUR FAILURES OF EMERGENCY POWER UNITS

2.3.1 Introduction

Four failures of emergency power units are discussed in Section 2.3. Emergency power units provide active safety as they start working upon a signal. The safeguarding or protection is not continuously present, and an activation is required. The four different failures of emergency power units to come into action have four different causes. The failure of active safety is in hospitals mostly followed up by procedural safety.

2.3.2 Twenteborg Hospital at Almelo in The Netherlands in 2002

On July 30, 2002, the Twenteborg hospital at Almelo in The Netherlands was struck by lightning [2, 3]. The external electric power supply was interrupted. In such a case, the emergency electric power supply should start automatically. Thus, this provision is an active safety measure. However, the diesel engines of the generators of the emergency power supply did not start because lightning had also damaged the circuitry of the emergency power supply. It took half an hour to repair the external electric power supply. Essential equipment was connected manually to a local accumulator in this period.

2.3.3 Westfries Gasthuis (Hospital) at Hoorn in The Netherlands in 2003

The external electric power supply of the Westfries Gasthuis (hospital) at Hoorn in The Netherlands was interrupted at 22.30 h on November 24, 2003 [4]. The emergency electric power supply should take over automatically in such a case. Similar to the previous case, this provision is an active safety measure. However, because of a faulty relay, the generators of the emergency electric power supply did not start. At 23.00 h, the fire brigade had installed emergency power supply generators for critical departments of the hospital. These departments were, e.g. intensive care, cardiology, and incubators. In the meantime, hospital personnel had taken care of the breathing upon of patients manually (procedural safety). Childbirths and operations did not take place at the time of the interruption of the external electric power supply. The external electric power supply had been fully restored at 03.30 h on November 25, 2003.

A notable aspect is that the emergency electric power supply did not work in spite of the fact that it had been successfully tested in October 2003.

2.3.4 ZGT Hengelo Hospital at Hengelo (O) in The Netherlands in 2011

The electric power supply to the ZGT Hengelo hospital at Hengelo (O) in The Netherlands was interrupted at 08.05 h on May 8, 2011 [5]. The cause was short-circuiting within the equipment controlling the power supply to the hospital. There was no interruption of the external power supply. The circuitry of the emergency power supply could not detect the interruption of the electric power supply to the hospital and hence did not activate the emergency electric power supply. The electric power supply to the hospital was restored provisionally by the supplier of the external electric power supply shortly past 09.00 h on May 8, 2011. Six patients were breathed upon in Intensive Care at the time of the power interruption. Partly by means of local accumulators and partly manually, the breathing upon of these patients could be continued. Two patients were transferred to a different hospital because they needed kidney dialysis. The supplier of the equipment controlling the power supply to the hospital repaired the short-circuiting in that piece of equipment in the course of May 8, 2011. The electric circuits were not modified.

2.3.5 Chemical Plant

A power failure occurred in a chemical plant. The activation of the emergency power unit was required to complete certain activities. However, the emergency power unit did not start up. On checking the situation, it appeared that the unit could not be activated as it had been switched off. A message had been attached to the diesel motor reading: "temporarily closed down." That measure had not been checked with the production staff.

2.3.6 Additional Remarks

In the first case, the Twenteborg hospital at Almelo in The Netherlands, the sequence of events was started by lightning. The immediate cause of the disturbance at the Westfries Gasthuis at Hoorn in The Netherlands was the interruption of the external electric power supply. Furthermore, the problems at ZGT at Hengelo (O) in The Netherlands started with a short-circuiting within hospital equipment. Finally, the emergency power unit in the chemical plant could not come into action due to a mistake. Thus, we see four different immediate causes of the problems.

2.4 THE FAILURE OF THE BLOWOUT PREVENTER (BOP) AT THE GULF OIL EXPLOSION IN 2010

An accident occurred on the Mobile Offshore Drilling Unit Deepwater Horizon in the Gulf of Mexico on April 20, 2010 [6, 7]. Control of the well was lost on the evening of that day, allowing hydrocarbons to enter the drill pipe and reach the drilling unit, which resulted in explosions and subsequent fires. Eleven crew members died, and others were seriously injured. The fires engulfed and ultimately destroyed the rig, which sank after approximately 36 h. The first of more than four million barrels of oil began gushing uncontrolled into the Gulf of Mexico on April 20, 2010. The flow from the well was stopped using a technique called "top kill" on July 20, 2010. The well was effectively dead after a relief well was completed and cement was pumped into the well to seal it. This was declared to be the case on September 19, 2010.

Regarding the cause, the first two conclusions of the National Commission on the Deepwater Horizon Oil Spill and Offshore Drilling [6] are quoted:

- The explosive loss of the Macondo well could have been prevented.
- The immediate cause of the Macondo well blowout can be traced to a series of identifiable mistakes made by BP, Halliburton, and Transocean that reveal such systematic failures in risk management that they place in doubt the safety culture of the entire industry.

The oil and gas industry began to move offshore in approximately 1960. The industry first moved into shallow waters and, as from approximately 1980, into deepwater where vast new reserves of oil and gas have been opened up. The Deepwater Horizon drilled the Macondo well under 5000 ft (1524 m) of Gulf water and then over 13 000 ft (3962 m) under the seabed to the reservoir below. The pressure in the water at seabed level is approximately 2250 psi (153 bar), and intervention at the seabed level is only possible by means of remotely operated vehicles (ROVs). The reservoir pressure is also high. The reservoir temperatures are exceeding 200 °F (93.3 °C). It is clear that risks exist if a well gets out of control.

The engineering and design of the well started in 2009. On April 9, 2010, the well was drilled to its final depth of 18 360 ft (5596 m).

In the event of a loss of well control, various components of the BOP stack are functioned in an attempt to seal the well and contain the situation (see Figure 2.1). The lower section of the BOP attaches to the subsea wellhead. Prior to, during, and following the accident, numerous attempts were made to control the well by activating or functioning various components of the BOP. However, these attempts were unsuccessful. At the time of the accident, the drill pipe was present in the wellbore. The portion of the drill pipe between the shearing blades of the blind shear rams (BSRs) of the BOP was off center and held in this position by buckling forces. The BSRs are the only set of rams designed to cut drill pipe and seal the well in the event of a blowout. Because the trapped portion of the drill pipe was off center, the BSRs could not cut the drill pipe.

Forensic investigations by Det Norske Veritas proved that the BSRs of the BOP had been activated [7]. It is stated in their Executive Summary:

"Of the means available to close the BSRs, evidence indicates that the activation of the BSRs occurred when the hydraulic plunger to the Autoshear valve was successfully cut on the morning of April 22, 2010. However, on the evidence available, closing of the BSRs through activation of the AMF/Deadman circuits cannot be ruled out."

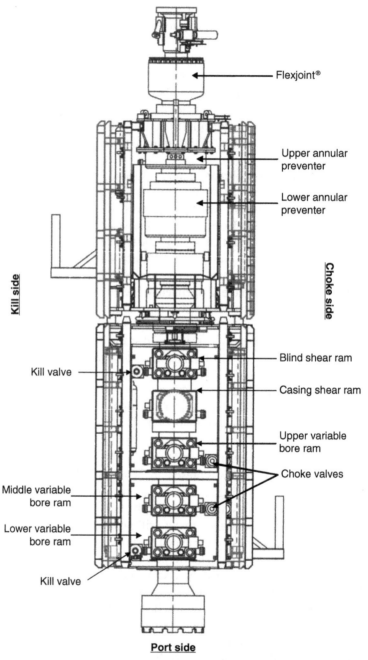

Figure 2.1   Deepwater Horizon BOP. *Source:* Courtesy of Bureau of Safety and Environmental Enforcement, Washington D.C., U.S.A.

AMF stands for automatic mode function. The Autoshear valve was cut by an ROV on the morning of April 22, 2012. If closing of the BSRs occurred through activation of the AMF/Deadman circuits, such closing would have occurred earlier than April 22, 2010, e.g. on April 20, 2010. Whether on April 20, 2010 or on April 22, 2010, activation of the BSRs did, as already stated, not lead to the sealing of the well.

The BSRs in the BOP of the Deepwater Horizon were an active safety measure. The safety measure did not function.

The BOP is a last line of defense against the loss of well control.

2.5 THE SAFEGUARDING OF FORMULA ONE RACES

The Brazilian Formula One driver Ayrton Senna died in a crash at Imola in Italy on May 1, 1994. Since that accident one further life was lost at Formula One races. A serious accident occurred at Suzaka in Japan on October 5, 2014. The French Formula One driver Jules Bianchi was heavily injured at this accident and died on July 17, 2015. Still, the situation improved considerably if one compares the number of accidents and incidents in the period 1994–2014 to that in the period 1974–1994.

First, passive safety measures for the driver will be mentioned. Possibly, the most important measure in this category is the head restraint Head And Neck Support (HANS) device. It had been realized that a number of fatalities were due to the unrestrained head leading to excessive loads to the neck and base of the skull at frontal impacts. The inventor is Ron Hubbard. HANS was introduced in 2002. From 2003, it became mandatory in Formula One races. HANS became mandatory in other branches of motor sport as well.

The incorporation of a strong cockpit designed to stay intact in the event of an accident is a further passive protection measure. The introduction of fuel tanks made of strong fibers also belongs to this category.

In the mid-1970s, FIA (Fédération Internationale de l'Automobile) introduced standards for clothing and helmets. Over the years, these standards have become increasingly strict. Suits, shoes, gloves, helmets, seats, and other accessories are now made from a fire-resistant material.

Furthermore, there is a five-point harness securing a driver to his seat. A quick-release mechanism enables a driver to get out of the car in an emergency.

Passive safety measures for the layout of the racing tracks are discussed briefly. FIA saw to the elimination of all dangerous locations. Wide strips adjacent to the racing track, gravel pits, speed-limiting chicanery, concrete walls, and adequate distances between the cars and the spectators made the Formula One races safer.

The automatic interruption of the fuel supply to the motor when an accident occurs is an active safety measure. There are further active protection measures.

Procedural measures have also been taken. Race control is informed by marshals, and the course of the race is checked on monitors by race control. The safety car has been introduced. The safety car slows down the race in the event of a crash or other incidents. The medical car is used to rush doctors and rescue personnel to a driver who is injured during a race.

2.6 DUST EXPLOSION RELIEF VENTING

The principle of dust explosion relief venting is that, at a predetermined pressure, an aperture opens to vent the explosion products safely from, e.g. a dust filter (see Figure 2.2). It is a passive safety measure. For a short period after the vent opens, the pressure may continue to rise, so sufficient area should be provided to ensure that the pressure peak does not damage the vessel. This method can be used only if the emission of material is allowable and a safe discharge area for the products can be found.

A design based on the invention of the Davy safety lamp in 1815 for use in coal mines is an interesting development. The lamp in use before 1815 had a wick and an oil vessel originally burning a heavy vegetable oil. Davy

Figure 2.2 Explosion vent. *Source:* Courtesy of Rembe GmbH Safety + Control, Brilon, Germany.

discovered that a flame enclosed inside a copper mesh of a certain fineness cannot ignite methane, the main component of flammable gases in mines. The minimum explosible concentration of methane in air is between 4% and 5% by volume. The screen cools the passing gases and thereby acts as a flame arrestor. The type of dust explosion relief vent based on the idea to cool emitted gases is marketed by several suppliers. It encompasses a discharge into a device having a cylindrical wall consisting of several layers of metal gauze that, in case of a dust explosion, cool emitted gases and retain solid particles. Be it that a larger vent size is necessary because the gauze reduces the opening. Thus, discharge into a plant building is possible (see Figures 2.3 and 2.4).

Figure 2.3   Dust explosion relief venting. *Source:* Courtesy of Stangl Reinigungstechnik GmbH, Strasswalchen, Austria.

Figure 2.4 Dust explosion relief venting detail. *Source:* Courtesy of Hoerbiger Ventilwerke GmbH & Co. KG, Vienna, Austria.

REFERENCES

[1] Kletz, T.A. and Amyotte, P.R. (2010). *Process Plants – A Handbook for Inherently Safer Design*, 6, 187–189. Boca Raton: CRC Press.

[2] De Twentsche Courant Tubantia, Enschede, The Netherlands, August 1, 2002, p. 10 (in Dutch).

[3] De Twentsche Courant Tubantia, Enschede, The Netherlands, August 7, 2002, p. 7 (in Dutch).

[4] NRC Handelsblad, Rotterdam, The Netherlands, November 25, 2003, p. 3 (in Dutch).

[5] De Twentsche Courant Tubantia, Enschede, The Netherlands, May 10, 2011, pp. 6, 7 (in Dutch).

[6] National Commission to the President on the Deepwater Horizon Oil Spill and Offshore Drilling (2011). *Deep Water – The Gulf Oil Disaster and the Future of Offshore Drilling*, vi–xii, 21–53. Washington: The Superintendent of Documents, U.S. Government Printing Office.

[7] Det Norske Veritas (2011). *Forensic Examination of Deepwater Horizon Blowout Preventer, Final Report*, vol. I, 1–17, 34, 35, 169, 174. Dublin, OH: Det Norske Veritas.

3

SAFETY IMPROVEMENTS
OVER THE YEARS

3.1 INTRODUCTION

The concept of intrinsic safeguarding has been discussed in Chapter 1. It is known that, in the past, already many steps have been taken in different fields to improve safety. A number of these steps are dealt with in this chapter. The improvements can, in several instances, be called an implementation of intrinsic protection. The measures were often taken after serious problems were encountered. A subdivision is made in this chapter between transport, industry, and society. Furthermore, regarding transport, a subdivision is made between transport by road, rail, sea, and air.

The cases discussed are meant as examples. They do not provide an overall picture of the fields considered.

3.2 TRANSPORT

3.2.1 Road Transport in The Netherlands

The safety of road transport in The Netherlands, like in many other countries, improved remarkably over the last 50 years [1]. More than

Safety in Design, First Edition. C.M. van 't Land.
© 2018 John Wiley & Sons, Inc. Published 2018 by John Wiley & Sons, Inc.

3000 people died in road traffic in The Netherlands annually in the 1970s. In 2010, the figure was 640. A gradual decrease in the numbers can be noticed between 2000 and 2010. In 2011, 661 people died in road traffic. The improvement is even greater on expressing the numbers per kilometer traveled because the mobility has increased over the period considered. The number of people seriously injured in road traffic, however, increased gradually from 15 424 to 20 100 between 2006 and 2011.

Three aspects are probably the most important ones concerning the safety of road transport: the design of the infrastructure, the vehicle safety, and the drivers' conduct.

The number of people who died per billion motor vehicle kilometers in The Netherlands in 2009 is 5.0. This figure is comparable to those of other developed countries.

As stated, 661 people died in road traffic in The Netherlands in 2011. That figure is made up of 231 occupants of motor vehicles, 200 cyclists, and 230 other road users.

3.2.2 Unidirectional Road Traffic in Tunnels

Modern tunnels for road traffic are built unidirectional. Thus, head-on collisions between vehicles are impossible. That type of collision has occurred in bidirectional tunnels. A head-on collision that has occurred in the Gotthard road tunnel in Switzerland in 2011 will be discussed shortly in this section and the Swiss Sierre tunnel will be mentioned as an example of a unidirectional tunnel in this section too.

The Gotthard road tunnel is a bidirectional tunnel and was opened in 1980. It is almost 17 km long. A collision of two trucks occurred on October 24, 2001, and caused a fire in the tunnel, killing 11 people and injuring a larger number of people. The smoke and the gases produced by the fire were the main causes of death. The effects of fires in tunnels are serious because gases and heat cannot disperse quickly and because the oxygen in the air is depleted rapidly. As the oxygen supply may be inadequate for complete combustion of combustible materials, carbon monoxide may be formed. That gas is toxic at low concentrations, e.g. a concentration of 0.4% by volume in air proves lethal in less than 1 h [2].

The Swiss Sierre tunnel is a unidirectional tunnel and was opened in 1999. It is 2.6 km long. In this type of tunnel, a head-on collision is, as stated, impossible. However, in 2012, still, a serious accident occurred in this tunnel, which will be mentioned in Section 6.2.1.

3.2.3 Rail Transport in The Netherlands

Heavy rail in The Netherlands will be considered only in this paragraph. Light rail, i.e. underground railway and trams, will not be discussed. The safety of rail transport improved considerably after an accident at Harmelen in 1962. It was the worst train accident in The Netherlands until now. It took 93 lives and 52 people were wounded [3].

The accident occurred as follows. A fast train was on its way from Leeuwarden to Rotterdam, and, just before Harmelen, had a speed of 125km h^{-1}. The driver missed a warning sign (yellow signal). At the place and the time of the accident, there was a dense fog. When seeing a warning sign, the driver must adapt the train's speed because the warning sign can be followed by a stop sign (red sign). However, he did not adapt the train's speed. As a matter of fact, the next sign was a stop sign. The driver then activated the brakes; however, he could not prevent the train from passing the stop sign and colliding, almost head-on, with a stopping train from Rotterdam to Amsterdam. When the collision took place, the express still had a speed of 107km h^{-1}.

After this accident, the Dutch government took the decision to accelerate the introduction of a safeguarding system. The Dutch acronym for this system is ATB. The objective of the system is to prevent trains from passing a stop sign (red sign). The introduction of the system took more than 30 years. In the course of the years, the original system has been further improved. The protection system informs the driver about the maximum allowable speed of the train. If the driver fails to adjust the speed, the safeguarding system brings the train to a halt. The system has two shortcomings. The first one is that the safeguarding system is not active when the train speed is lower than 40km h^{-1}. The reasoning behind this shortcoming is as follows. If the driver has already, on approaching a stop sign, adjusted the train's speed to a speed lower than 40km h^{-1}, it is highly unlikely that he will subsequently pass the stop sign. The second shortcoming is related to the brake-path. A stop sign (red sign) is preceded by a warning sign (yellow). When the driver misses a warning sign, the system takes over. However, it is not always possible for the system to bring the train to a complete halt before the stop sign. Because of these two shortcomings, trains still do pass stop signs and that sometimes leads to accidents. In 2010, the number of times a train passed a stop sign was 169. Between 2001 and 2009, the annual number was greater than 200 [4]. The number has the tendency to decrease.

In 2013, it was more than 20 years ago that a rail accident occurred at which the lives of more than one passenger were taken. This figure does not include accidents at railroad crossings. In 2011, nine lives were lost due to accidents on railroad crossings. The annual number of suicides on the railroad amounts to approximately 200 [4].

3.2.4   Chlorine Transport by Rail

Akzo Nobel transported liquid chlorine by rail between plants in The Netherlands in the past. Those transports stopped in 2006. The production of chlorine by the company did not stop; however, the chemical was, as from 2006, produced and converted into other chemicals at the same site. The production of monochloroacetic acid (MCA) from acetic acid and chlorine is a typical example. Most of the MCA is used to manufacture several hundred thousand tons annually of carboxymethyl cellulose (CMC). Starch can be reacted with MCA to give carboxymethyl starch, which is as widely used as CMC. Another major application of MCA is the production of herbicides based on aryl hydroxyacetic acids.

Chlorine is very toxic [5]. The boiling point of liquid chlorine is −34.05 °C at atmospheric pressure. The saturated vapor pressure at 30 °C is about 9 bar. Thus, if liquid chlorine is not cooled, it can only be transported under pressure. If liquid chlorine is at a temperature of 30 °C and the pressure is suddenly reduced to atmospheric pressure unexpectedly, 20% of the liquid chlorine evaporates rapidly. In other words, it is then a flashing liquid. The remaining 80% then cools down to −34.05 °C and subsequently evaporates more slowly. The density of gaseous chlorine relative to air is 2.48. Thus, gaseous chlorine stays at ground level and can be displaced by the wind.

As stated, liquid chlorine is no longer routinely transported by rail in The Netherlands. Incidentally, those transports still occur. For example, Akzo Nobel buys chlorine in Germany and has it transported to Rotterdam when the chlorine plant at Rotterdam is maintained. Chlorine is transported under pressure until now. The possibility of transporting cooled, pressureless liquid chlorine by rail was discussed before 2006.

3.2.5   Sinking of the RMS Titanic in 1912

The RMS (Royal Mail Ship/Steamer) Titanic sank in the Atlantic Ocean on April 15, 1912, due to collision with an iceberg. The estimated number of people on board was 2224. This number concerns both passengers and crew. The number of people saved was 710 and the number of people having lost their life was 1514.

This accident has been investigated thoroughly and important changes to maritime regulations were recommended. This gave rise to the establishment of the International Convention for the Safety of Life at Sea (SOLAS), which is still in existence today. This Convention harmonizes maritime safety regulations internationally. It is mentioned in this context that it was agreed that radio equipment on passenger ships should be manned around the clock. The background is that, at the time of the calamity, the radio equipment of the nearby SS (Steam Ship) California was not manned. Thus, it could not receive the radio signals of the RMS Titanic. If the radio equipment of the SS California had been manned, hundreds of lives might have been saved. Furthermore, the International Ice Patrol was founded in response to the sinking of the RMS Titanic. This organization has the task of monitoring the presence of icebergs in the Atlantic and Arctic Oceans. Their movements are reported for safety reasons. Vessels that have, since 1913, heeded the Ice Patrol's published iceberg limit have not collided with an iceberg.

3.2.6 Oil Tankers with Double Hull

The United States, the European Union, and other countries have phased out or are phasing out single-hulled tankers. In tankers of a single-hull design, the oil in the cargo tanks is separated from the seawater by one metal wall only. In tankers of a double-hull design, the metal wall containing the oil is protected against damage by a second metal wall at a sufficient distance from the inner wall. Two accidents with oil tankers having a single hull are recapitulated shortly.

The first accident is the Exxon Valdez oil spill [6]. The Exxon Valdez, carrying crude oil, was on its way from a pipeline terminal at Valdez, Alaska, to Long Beach, California. On March 24, 1989, it ran aground at Prince William Sound, Alaska. In order to avoid ice, the ship proceeded outside the tanker lane. The spill's volume has been estimated at $41\,000\,\mathrm{m}^3$. The region where the ship stranded is a habitat for salmon, sea otters, seals, and seabirds. Much damage was done to the habitat. A US Coast Guard study undertaken after the accident indicated that up to 60% less oil would have entered the water if the Exxon Valdez had been equipped with a double hull.

The second accident concerns the Prestige oil spill [7]. The Prestige was on its way from St. Petersburg, Russia, to Gibraltar. It contained 77 000 tons of heavy fuel oil. It suffered severe damage during a storm off Galicia, in north-western Spain, on November 13, 2002. The ship was 26 years old. The French, Spanish, and Portuguese governments denied the ship access to their ports. On November 19, 2002, the ship broke into two and sank in deep waters off the Galician coast. It is estimated that more than 80% of the ship's cargo has been spilled off Spain's north-east coast. Marine life was strongly affected by the accident. It is not certain whether a double hull, with all other things being equal, would have prevented the spill.

Ships with a double hull have a second line of defense. The oil tankers with a single hull are being phased out because the release of oil generally does much damage to the environment.

3.2.7 Two Comet Accidents in 1954

Event Number 1 An aircraft of the type Comet 1 and operated by British Overseas Airways Corporation (BOAC) left the Ciampino Airport at Rome at 09.31h on January 10, 1954, on a flight to London [8]. At approximately 09.51h, the communication between the captain of a different airplane and the Comet 1 was suddenly interrupted. The aircraft was at this time probably approaching a height of 27 000 ft (8230 m). Four witnesses from Elba saw the airplane crash into the Mediterranean Sea at about 10.00h. Part of the Comet 1 fell into the sea in flames. Something had happened to the aircraft with catastrophic suddenness. A total of 29 passengers and 6 crew members lost their lives in the accident.

Event Number 2 An airplane of the type Comet 1 and operated by South African Airways left the Ciampino Airport at Rome at 18.32h on April 8, 1954, on a flight to Cairo [8]. Something catastrophic happened to the aircraft over the Mediterranean Sea near Naples when it must have been at or near the end of the climb to 35 000 ft (10 668 m) at about 19.10h. A total of 14 passengers and 7 crew members lost their lives in the accident.

Comet History de Havilland Aircraft Company had built only military aircraft during the Second World War. In 1945, they wanted to resume the manufacture of civil aircraft. As they had several years' experience with jet fighters, the production of civil jet aircraft was attained. The first civil aircraft with jet engines was the Comet 1. Passenger service started in May 1952. A Comet 1 aircraft is depicted in Figure 3.1.

Comet Characteristics The Comet 1 had four jet engines giving the aircraft a cruising speed of 400 miles per hour (644 km h^{-1}). The engines were integrated into the wings. This speed was 1.5 times the speed of a DC-6 airplane. It was essential that the cruising height should be upward of 35 000 ft (10 668 m) – double that of the airliners the Comet 1 competed with. The atmospheric pressure is 3.5 psi (0.24 bar) at an altitude of 35 000 ft (10 668 m), whereas the atmospheric pressure is 7.8 psi (0.53 bar) at an

Figure 3.1 A Comet 1 aircraft. *Source:* Courtesy of Charles Brown Collection, RAF Museum, London, UK.

altitude of 17 500 ft (5334 m). The pressure within the fuselage is usually maintained at 75–80% of the atmospheric pressure at ground level. Thus, the difference between the internal and the external pressure was 8.25 psi (0.56 bar) for the Comet 1 and 4.0 psi (0.27 bar) for airplanes the Comet 1 competed with.

The firm de Havilland used a design pressure of 2.5 times 8.25 psi (0.56 bar). The cabin was tested at two times 8.25 psi (0.56 bar). It was believed that a cabin that would survive undamaged a test at double its working pressure would not fail in service under the action of fatigue due to the pressurization to working pressure on each flight and to other fluctuating loads to which it was subjected in operations.

It would have been possible to make the cabin stronger; however, that would have meant a thicker cabin wall and more mass for the aircraft.

The name fatigue suggests a weakness of metals and other materials causing them to break under a load that is repeatedly applied and then removed, though they can support a much larger load without distress.

The Cause of the Accidents The Abell Committee investigated the cause of Event Number 1. Several improvements were recommended and implemented. The Comet fleet of BOAC was grounded after Event Number 2. Event Number 2 was investigated by the Royal Aircraft Establishment (RAE) led by Sir Arnold Hall. It struck RAE that there had been two accidents in what appeared to be similar conditions, each occurring at about the time when the aircraft was nearing the top of its climb. Their attention became directed to fatigue because of the circumstances of the accidents and because of the fact that the modifications carried out after Event Number 1 seemed to rule out many of the other possible causes. It was decided to make a repeated loading test of a whole cabin. The cabin was filled with water, and more water was pumped in until the desired difference between the internal and the external pressure (8.25 psi, 0.56 bar) was reached. The whole cabin was immersed in a tank, and the tank and the cabin were simultaneously filled with water. Water was used rather than air because, in case of a leak, the pressure will be relieved immediately because water is incompressible. The object of the tests was to simulate the conditions of a series of pressurized flights. Moreover, the program of tests included, at intervals of approximately 1000 "flights", a proving test in which the pressure difference was raised to 11 psi (0.75 bar). A proving test is a test to check that the cabin does not exhibit a permanent deformation when the selected pressure difference is applied. The cabin had made 1230 pressurized flights before the test and after the equivalent of a further 1830 such flights, making a total of 3060, the cabin structure failed, the starting point being the corner of one of the cabin windows. It was suggested by Dr. Walker of RAE that the "life" of 3060 flights in the test might be equivalent to about 2500 real flights.

An inspection of the wreckage recovered after Event Number 1 revealed that the first fracture of the cabin structure occurred near a rear ADF (Automatic Direction Finding) window and spread fore and aft from it. There were two of these windows located at the top of the cabin structure.

The opinion of RAE was that Event Number 1 was caused by structural failure of the pressure cabin, brought about by fatigue. The conclusion was based on both the repeated loading test and inspection of the wreckage. As to Event Number 2, owing to the absence of wreckage, RAE was unable to form a definite opinion on the cause of the accident. They drew the attention to the fact that the explanation offered for Event Number 1 appeared to be applicable to Event Number 2.

de Havilland had, in 1953, carried out repeated loading tests. It was concluded on the basis of these tests that there was, concerning the safety of the Comet's cabin, an ample margin. 8.25 psi (0.56 bar) was applied 18 000 times. These tests were ended by a failure of the skin in fatigue at the corner of a window, originating at a small defect in the skin. However, in retrospect, these tests were not representative. First, they had used a test section representing the nose section of the cabin and not the complete cabin. Second, the specimen had been subjected to some 30 earlier applications of pressures between 8.25 psi (0.56 bar) and 16.5 psi (1.12 bar). If a part of a cabin is statically tested in this manner, some regions will be plastically deformed. If, subsequently, this part of a cabin is submitted to fatigue tests, it will stand the tests relatively well. That is, better than a part of a cabin that has not been subjected to such high pressures and is used for fatigue tests. de Havilland admitted this effect; however, they considered the number 18 000 so large that the results of the fatigue tests could be accepted. A possible number of pressurizations per annum is 600, so that the number 18 000 corresponds to 30 years. This number of years is much larger than the expected lifetime of the airplane.

The Court of Inquiry of the Ministry of Transport and Civil Aviation concluded that de Havilland was, regarding the design of the Comet, proceeding in accordance with what was then regarded as good engineering practice.

At the time of the Elba accident, the Comet concerned had made 1290 pressurized flights. At the time of the Naples accident, the Comet concerned had made 900 pressurized flights. The cabin tested by RAE failed after 3060 pressurized flights. These figures are in the same range of accuracy.

Additional Remarks There had been accidents with the Comet 1 aircraft prior to the two accidents described. Passenger service with an improved version, the Comet 4 airplane, was resumed in 1958.

Concluding Remarks The Comet 1 was the first civil aircraft equipped with jet engines. Compared to the DC-6, equipped with propeller engines, the Comet 1 flew 1.5 times faster at the double altitude. This fact implied that the difference between the internal and the external pressure was, for the Comet 1, double the difference for the DC-6. Important lessons, especially concerning the phenomenon fatigue, were learned from the two Comet accidents. de Havilland's design rule that a cabin having a design pressure of 2.5 times the difference between the internal and the external pressure and surviving undamaged a test at double this difference would not fail under the action of fatigue due to the pressurization to working pressure on each flight, and to other fluctuating loads to which it was subjected in operations, appeared to be incorrect for the Comet 1 aircraft.

In this respect, it is possibly of interest to note the design load of elevator cables in The Netherlands. Elevator cables are designed to break at not less than 12 times the normal maximum load.

3.2.8   Helium Gas for Zeppelins – Zeppelin Crash in 1937

On May 6, 1937, the German airship Hindenburg crashed at Lakehurst Naval Air Station in Lakehurst Borough, New Jersey, US. It took fire while trying to dock with a mooring-mast. There were 97 people on board (36 passengers and 61 crewmen), of which 36 died. One groundsman also died. The airship could take fire because it was filled with hydrogen, an inflammable gas. Hydrogen is very easy to ignite; it has low ignition energy. The cause of the ignition has not been established with certainty although hypotheses have been put forward. It is likely that a spark due to electrostatic discharge was the cause of the ignition. American airships had helium, a noninflammable material, as carrier gas.

Helium is a much better option than hydrogen because its choice removes at least one potential cause of an accident. Still, serious accidents also occurred with helium-filled airships.

There was a keen competition between airships (Zeppelins) and airplanes in the first half of the previous century. The Hindenburg accident marked the end of the airship era. In 1993, an initiative was taken to again start with the building and exploitation of commercial airships. The German company ZLT Luftschifftechnik, having its base at Friedrichshafen in Germany, started the project in 1993. The maiden flight of the Zeppelin NT (New Technology) was on September 18, 1997. Four years later, the Deutsche Zeppelin-Reederei (DZR) recommenced commercial air traffic. The lift for these airships is provided by helium.

3.3   INDUSTRY

3.3.1   Cotton Spinning Plants

There was textile industry in the eastern part of The Netherlands between 1820 and 1970 [9]. Cotton mills were an important part of this industry. In these mills, cotton was spun into threads, an activity that generates a lot of inflammable dust. Typically, the buildings had four storeys. Wood was used as an important material of construction for the first generation of these cotton mills, built between 1820 and 1900. Thus, when dust was ignited, the building could take fire and burn down. As a matter of fact, one out of three to four cotton mills actually burnt down in those years. As from 1900, wood as a material of construction was replaced by cast iron, steel, and concrete. Furthermore, automatically closing doors were installed to prevent the spreading of a fire. Typically, cotton mills constructed after 1900 had a tower containing staircases with a water basin at the top

Figure 3.2   Cotton mill with water tower. *Source:* Courtesy of De Museumfabriek, Enschede, The Netherlands.

(see Figure 3.2). In case of fire, water could be released from that basin via pipes attached to the ceilings to extinguish a fire. The first generation of these so-called sprinkler installations were hand-operated, whereas the second generation released the water automatically when solder in the water tubes melted when heated. Cotton mills were among the first industrial buildings equipped with sprinkler installations. Furthermore, cotton mills typically had a dust extraction system that passed the dust on to a dust tower.

3.3.2   Akzo Nobel Extracts Salt Without Subsidence

Akzo Nobel Industrial Chemicals extracts salt from the soil in both the eastern and northern parts of The Netherlands. This paragraph concerns extraction in the eastern part of The Netherlands in a region called Twente [10]. Specifically, the area around the towns of Enschede and Hengelo (O) is in focus. The salt is extracted by means of solution mining, water is passed down into the salt layer, and the saturated brine flows to the salt plant located at Hengelo (O). Approximately 250 million years ago, the salt layer has been formed by the evaporation of water from the sea. The salt layer is 400–500 m below the surface and is approximately 50 m thick. The extraction of salt from the specific area started in 1933. Until approximately 1980, an approach was followed that would prevent subsidence at the surface. However, on several occasions, small subsidences did occur. In 1991, a large subsidence occurred and a sinkhole was created. This occurrence gave rise to a reconsideration, and it was decided to follow a new approach. This approach comprises two safeguarding measures in series.

The first protection measure comprises a salt solution method at which at least the upper 5 m of the salt layer are left intact. The upper 5 m of the salt layer are called the roof. With this thickness, it is modeled that the roof is strong enough to prevent a collapse of the roof.

The second protection measure considers the situation if, unforeseen, a collapse of the roof having a thickness of 5 m nevertheless occurs. The reasoning is as follows. If the roof breaks down, salt from the roof and materials of the layers between the salt layer and the surface will fall into the space below the roof. However, the materials falling down will rearrange themselves. In doing so, they will occupy a larger volume than the original volume. The larger volume should guarantee that an improbable collapse of the roof should not be noticeable at the surface.

If the second safeguarding measure cannot be achieved with a roof thickness of 5 m, the roof thickness is increased to ensure that a collapse of the roof will not be noticeable at the surface. In practice, this means that most of the time the roof thickness is increased to 40–50% of the salt layer thickness.

3.3.3   Two New Cocoa Warehouses at Amsterdam in 2011

In 2003, there were two large fires in warehouses where cocoa powder was stored north of Amsterdam in The Netherlands [11, 12]. It took the fire brigade 5 days to extinguish the first fire and 8 days to extinguish the second one. Cocoa powder is a difficult product to extinguish when it has taken fire. Important aspects are that the product contains 10–20% by weight of fat and that the powder melts when the temperature is increased. The only way to stop a fire is to spread the product. Furthermore, cocoa powder is a relatively expensive product.

In 2011, Cargill Cocoa and Chocolate planned two new warehouses for cocoa powder and other cocoa products west of Amsterdam. For both warehouses, one floor having an area of 10 000 m² was envisaged. DSV Solutions at Amsterdam was asked to supervise the construction of the warehouses. That company proposed to reduce the oxygen concentration in the air in the warehouses from 21% by volume to 16–17% by volume. Experiments carried out by TNO in The Netherlands showed that the cocoa products cannot take fire at the reduced oxygen level. The reduced oxygen level is maintained by monitoring the oxygen level in the warehouses and, if need be, supplying nitrogen gas automatically from an external supply. The warehouses are airtight.

Cocoa products can, however, still smolder at the reduced oxygen level. Experiments carried out by TNO showed that smoldering stops when the oxygen level is reduced to 12% by volume by the supply of an inert gas. If smoldering is detected by instruments, carbon dioxide is led automatically into the warehouses. Carbon dioxide has been chosen for this application because it can be stored as a liquid occupying a relatively small volume.

The two warehouses have been taken into use in 2012. Generally speaking, the reduced oxygen level of 16–17% by volume is harmless for people. However, it is, in principle, not envisaged that operators enter the warehouses. The pallets with products are transported into and out of the warehouses by automatically guided vehicles (AGVs). Working with AGVs is considered safer than working with manned fork-lift trucks. The experiences in the first two years of operation are, generally speaking, good. The new warehouses, because of the reduced oxygen level, airtightness, and AGVs, require more attention and precautions than conventional warehouses.

3.3.4 Flame Retardants

Flame retardants are chemicals that reduce the inflammability of polymers [13]. The rate of progress of a fire is substantially decreased, and, in addition, smoke development is delayed. However, flame retardants do not prevent polymers to take fire. Flame retardants started to be used on a large scale when polymeric materials replaced traditional materials such as wood and metals. This occurred in the 1970s. The new materials of construction were more combustible than the materials they replaced. Flame retardants greatly reduce the risks of fires on using polymeric materials. Television sets, carpets, and curtains are mentioned as examples of objects made from polymeric materials. As to the mechanisms, four different types can be distinguished:

1. Dilution: For example, the addition of clays to polymer systems (e.g. 50–200 parts by weight per 100 parts of polymer) reduces inflammability.
2. Generation of noncombustible gas: For example, aluminum trihydrate decomposes into aluminum oxide and water vapor at 230 °C. Typically, 50–100 parts by weight are added to 100 parts by weight of polymer.

3. Free radical inhibition: Many flame retardants consist of compounds containing bromine. These compounds generate bromic acid vapor in a fire that combines, also in the fire, with free radicals. The effect is a decrease in the rate of combustion. The free radicals are generated by the burning polymer. Some flame retardants contain chlorine instead of bromine. In this context, it is interesting to note that PVC has inherently good flame-retardant characteristics because of the high chlorine content.

4. Solid-phase char formation: Flame retardants acting according to this mechanism form insulating or minimally combustible chars on polymer surfaces. The char reduces volatilization of active fragments. For example, organophosphates are used as flame retardants for polyphenylene ethers and polyurethanes. Many polyphenylene ethers are engineering materials, whereas polyurethanes are used for highly elastic foams (mattresses, cushions, car seats), rigid foam (insulation mats), and rigid and flexible moldings such as steering wheels for cars.

One of the flame retardants produced by Akzo Nobel Chemicals in the past was triphenyl phosphate; it has a melting point of 48–50 °C. Flame retardants prolong the time available to escape from, e.g. a building. A further aspect is that the smoke development of a fire is strongly delayed. Most victims of a fire suffered from loss of sight with subsequent suffocation. Flame retardants that contain bromine or chlorine are being discussed at the moment because, in case of a fire, vapors that contain bromine or chlorine are generated.

3.3.5 Clamp-on Ultrasonic Flow Measurement

Measuring the rate of flow of fluids in lines is important in many industries, e.g. in the chemical industry, power stations, and waste-water treatment plants. Two groups of flow meters can be distinguished. The first group consists of meters in which a signal, representing the magnitude of the flow, is generated from the energy of the flowing fluid. They are called fluid-energy-activated flow meters. The second group of flow meters comprises meters deriving a signal from the interaction of the flow and an external stimulus. They are called external stimulus flow meters.

Clamp-on ultrasonic flow meters belong to the second category. It is a special feature of this method that a physical contact between the measurement and the fluid processed is not necessary. Ultrasonic waves are passed through the pipe wall and the fluid and are received by external sensors.

There are two different types of ultrasonic flow meters. Transit-time flow meters use two combinations of an ultrasonic transmitter and a receiver separated by a known distance. The difference in transit time between a signal traveling with the flow and a signal traveling against the flow is a measure of the fluid velocity. The product of the velocity and the cross-sectional area is the volumetric flow. The volumetric flow can be converted into a mass flow when the temperature of the flow and the specific mass of the fluid as a function of temperature are known (see Figure 3.3).

The second type of ultrasonic flow meter is the Doppler flow meter. It sends an ultrasonic beam into the flow and measures the frequency shift of reflections from discontinuities in the flow. This type of flow meter is suitable for suspensions and liquids containing gas bubbles.

The transit-time flow meter is met more often than the Doppler flow meter. Both meter types can be used for liquids and gases. However, the meter types are applied more often to liquids than to gases. The remainder of the discussion will be devoted to the transit-time flow meter and its application to liquids.

Figure 3.3   Clamp-on ultrasonic flow meter. *Source:* Courtesy of Flexim GmbH, Berlin, Germany.

It is often still possible to use a transit-time flow meter when the specific mass of the liquid varies. This is the case when the specific mass can be derived from the velocity of sound through the medium. The transit-time flow meter then also measures the velocity of sound in the medium. Temperature variations can be coped with. The temperature of the liquid is measured, and the appropriate relationship between specific mass and velocity of sound is selected.

Clamp-on ultrasonic flow meters cannot cause leakages because there is no physical contact between the fluid and the instrument. Thus, they are safer than flow meters which are in contact with the process flow. Furthermore, they cannot suffer from corrosion or erosion and do not cause pressure loss. A further aspect is that they can relatively easily be installed in an existing plant. It is an attractive feature that they can be used for lines having a large diameter, e.g. up to 1.5 m.

It is a drawback that ultrasonic flow meters measure velocity and not mass flow. Flow meters exist that measure mass flow directly, e.g. Coriolis flow meters. Coriolis flow meters, however, are in contact with the process flow.

An example of the application of a clamp-on ultrasonic flow meter will be given [14]. It concerns the feed line of a concentration unit for dilute sulfuric acid having a temperature of up to 115 °C and containing metal salt particles. The line material is glass-fiber-reinforced plastic with an internal lining of the copolymer of hexafluoropropylene and tetrafluoroethylene (FEP). The line diameter exceeds 60 cm, whereas the wall thickness is 1 cm. It is reported that the instrument functions reliably and trouble-free as from June 2005.

3.4 SOCIETY

3.4.1 Inundation of Part of The Netherlands in 1953

Event   Large parts of the south-west part of The Netherlands were flooded on February 1, 1953 [15]. The water in the North Sea was driven up by a very heavy storm, at times even a hurricane, blowing from a north-westerly direction. This caused a storm surge. It was spring-tide when the storm blew. Many dikes could not deal with the resulting high water level due to this coincidence and broke. 1835 people drowned, 47 300 buildings were destroyed or seriously damaged, and approximately 153 000 hectares were inundated.

Previous History   It was known that many dikes in the south-western part of The Netherlands were too low at the time of the flood of 1953.

J.A. Ringers wrote in a memorial volume on the occasion of the 50th anniversary of Queen Wilhelmina's reign in 1948 a contribution titled "Department for the maintenance of dikes, roads, bridges, and the navigability of canals." The sentence "The dikes cry out for elevation" appears in this contribution [16]. A second sentence appears in his contribution: "The situation near Dordrecht cannot stay as it is now" [16].

Ringers had been the Director-General of the Department for the maintenance of dikes, roads, bridges, and the navigability of canals of the Dutch Ministry of Infrastructure and the Environment from 1930 to 1935. He had been a minister of The Kingdom of The Netherlands in the years 1945 and 1946.

The reason why the Dordrecht situation was mentioned specifically by him is as follows [17]. The main water dam of the Isle of Dordrecht had a height of 3.00 m above Normal Amsterdam Level (NAP). The height could be raised to 3.25 m above NAP by means of flood planks. The water reached a height of 3.43 m above NAP during the storm surge of 1916. It was neap-tide at the time of this storm surge. It is hence not too bold to surmise that the water would have come 53 cm higher if the storm surge would have occurred 1 week earlier (at spring-tide). On adding effects like the need to elevate the dikes because of the descent of the soil, the elevation due to normal tide, and the elevation due to civil structures at the shores and at the banks of rivers and canals, the water could rise to 4.50 m above NAP. The soil of The Netherlands descends with 20 cm per century [16]. And even the figure 4.50 m above NAP could, according to J. van Veen, only be considered as a "tentative minimum" as even more forces could exert an influence. Van Veen was an engineer in the Department for the maintenance of dikes, roads, bridges, and the navigability of canals in The Netherlands. As a matter of fact, the water inundated the center of Dordrecht in 1953.

Follow-up A plan was made to prevent inundations of the south-western part of The Netherlands in the future. The plan is known as the Deltaplan, and it has been executed. It comprises, among other things, cutting off several sea arms by means of dams or barriers that can, if need be, be closed, and raise the elevation of the dikes to a safe level. The design of the dikes, dams, and barriers is based on calculated values for the failure rates of the provisions. A typical example of a failure rate will be given. The connection between Rotterdam in The Netherlands and the North Sea is called Nieuwe Waterweg (New Waterway). This connection can be closed at the seaside by means of a barrier if need be. The barrier is called Maeslantkering (see Figure 3.4). It consists of two arms that can swing

Figure 3.4 Dutch sea barrier (Maeslantkering). *Source:* Courtesy of Department for the maintenance of dikes, roads, bridges, and the navigability of canals of the Ministry of Infrastructure and the Environment, The Hague, The Netherlands.

from the shores and meet each other in the middle of the canal. The failure rate of this provision is once per 10 000 years [18]. The meaning of this figure is that it is expected to occur once per 10 000 years that, when the Nieuwe Waterweg has been closed, the provision is too low. There is a further aspect. The Maeslantkering is an active safeguarding step. That means that there is a certain probability that the Maeslantkering cannot be closed although closing is required and activated. That probability is approximately 0.01.

3.4.2 Replacement of Coal Gas by Natural Gas in The Netherlands

Coal gas is an inflammable gaseous fuel obtained by heating coal in the absence of air. It was the primary source of gaseous fuel until the introduction of natural gas. It is also called town gas. The replacement of town gas by natural gas in The Netherlands took place after the discovery of natural gas resources in the Northern part of the country in 1959. Similar developments occurred in other countries. Town gas contains approximately 8% by volume of carbon monoxide. Carbon monoxide is a toxic gas; a concentration of 0.4% by volume in air is fatal for humans in exposures of less than

1 h [2]. Carbon monoxide is not a poison; it is a chemical asphyxiant producing a toxic action by combining with the hemoglobin of the blood. Small amounts of carbon monoxide in the air cause toxic reactions to occur because the affinity of carbon monoxide for hemoglobin is 200–300 times that of oxygen. Over the years, town gas has caused many fatal carbon monoxide poisonings. An important part of the deaths concerned suicides.

Besides carbon monoxide, town gas contains hydrogen and methane mainly.

Natural gas is safer than town gas because it does not contain toxic components. It can cause a gas explosion; however, town gas can also cause a gas explosion. Incomplete combustion of natural gas can lead to the formation of carbon monoxide.

3.4.3 CFCs

CFCs stand for chlorofluorocarbons [19]; these compounds consist of the elements chlorine, fluorine, and carbon. Their first use was in the 1930s as refrigerants. The reason for their application was as follows. Before 1930, ammonia, sulfur dioxide, methyl chloride, propane, and butane were used as refrigerants. The first two of these compounds are toxic, whereas the other materials are inflammable. Because of these properties, a number of accidents happened. A particularly serious accident happened in a hospital at Cleveland in the United States on May 15, 1928. A leak in the hospital's methyl chloride refrigeration system caused an explosion and a fire in which 128 people died. A combination of companies decided to develop an alternative.

CFCs were the alternative, they are nontoxic, noninflammable, and have the right thermodynamic properties for refrigeration. Their introduction was successful.

In the 1950s, CFCs were introduced successfully as spray can propellants. Their physical properties made them suitable for this application. It appeared that CFCs could also successfully be used for blowing polymer foams.

In the 1970s and 1980s, it was discovered that CFCs attack the ozone in the stratosphere. Ozone absorbs harmful UV-B radiation contained in sunlight. An increase in the number of skin cancer cases would result if more of this radiation would be present at ground level. In 1987, the Montreal Protocol, restricting the production and use of CFCs, was signed by 27 nations. The London amendments, issued in 1990, comprised a total ban on CFCs by the end of the twentieth century.

For refrigeration, CFCs were replaced by HFCs. Like CFCs, HFCs are nontoxic, are noninflammable, and have suitable properties for refrigeration. These compounds contain the elements hydrogen, fluorine, and carbon. In contrast to CFCs, they do not contain chlorine. HFCs do not deplete the ozone layer in the stratosphere.

Inflammable materials, such as dimethyl ether, have replaced CFCs as spray can propellants. HFCs could not replace CFCs for this application because they are greenhouse gases. Greenhouse gases are gases that, when present in the atmosphere, cause global warming. HFCs are at least 1000 times more potent greenhouse gases than CO_2. Different from the application as refrigerants, their use in this case would imply a deliberate emission into the atmosphere.

Materials such as carbon dioxide and cyclopentane have replaced CFCs as blowing agents for polymer foams. HFCs could not replace CFCs either in this case.

3.4.4 Dioxin in Feed

Too high dioxin levels were found in pork, poultry meat, and eggs in Germany in 2011 [20]. It appeared that fat sold by a company to a producer of cattle feed and poultry feed contained too much dioxin. The cause was that, within the premises of the latter company, vegetable fat had been mixed with fatty acid originating from the manufacture of biodiesel fuel. The dioxin levels of fatty acid originating from the manufacture of biodiesel fuel are too high and it is unsuitable for the production of cattle and poultry feed. Germany took subsequently the initiative to propose to the European Union a measure concerning companies that process both vegetable fats and so-called technical fats. The latter fats can contain too high dioxin levels. The measure is that the production lines of vegetable fats and "technical" fats should be separated to avoid contamination. The measure applies to both storage and production equipment. Germany's proposal was accepted by the European Union [21]. The implementation of the measure took a period of less than 2 years, whereas it could normally have taken more time. Other steps and measures to prevent the presence of dioxin in food and feed were also implemented, e.g. obligatory chemical analyses.

3.4.5 Street Motor Races in The Netherlands

"Road racing" means a course on closed public road. A large number of these races were organized in the past; however, few races have survived. In The Netherlands today, two races on street circuits exist, that is, at Hengelo in the province of Gelderland and at Oss. The history of the international races at Tubbergen in The Netherlands will be reviewed shortly [22].

The circuit had the form of a triangle with the villages of Tubbergen, Fleringen, and Albergen at the angular points. It had a length of almost

10 km and had 27 bends. The races were organized between 1946 and 1984. Between 1946 and 1972, 5 persons lost their lives at the races. The number of five casualties includes a racing motorist and a spectator in 1972. The races were not organized in 1973 and 1974. A restart occurred in 1975, and the last races on the original circuit took place in 1981. The last race at Tubbergen on a different circuit occurred in 1984.

3.4.6 An Unexpected Effect: Squatters Wear Moped Safety Helmets

Mr. Marcel van Dam was Parliamentary Under-Secretary and Minister in Dutch Cabinets in the 1970s and 1980s. He is quoted in the Dutch newspaper de Volkskrant as follows [23]: "We became conscious that many young people received brain damage at moped accidents. The Cabinet then decided to make the safety helmets obligatory. Thereupon the squatters discovered the helmet. The helmet made them unrecognizable and protected them against blows from the police. That is an effect, a change of a change, you see?"

The introduction of the protection method implied adverse effects in a different field.

REFERENCES

[1] Dutch Safety Board (2013). *Safety in Perspective*, 64–71. The Hague: Dutch Safety Board (in Dutch).

[2] Wiley Online Library (1999–2013). Carbon Monoxide. Ullmann's Encyclopedia of Industrial Chemistry. New York: Wiley Online Library.

[3] van Vollenhove, P. (2012). *Is It Unsafe Here? Impossible!* 149. Amsterdam: Uitgeverij Balans (in Dutch).

[4] Dutch Safety Board (2013). *Safety in Perspective*, 55–62. Dutch Safety Board, The Hague, The Netherlands (in Dutch).

[5] Wiley Online Library (1999–2013). Chlorine. Ullmann's Encyclopedia of Industrial Chemistry. New York: Wiley Online Library.

[6] Alaska Oil Spill Commission (1990). *SPILL-the Wreck of the Exxon Valdez-Implications for Safe Transportation of Oil. Final Report*. State of Alaska, Anchorage, pp. iii, 148.

[7] Harrington, J. (2003). *The Prestige Oil Spill Disaster and its Implications*, 1–5. Geneva, Switzerland: The Center for International Environmental Law.

[8] Cohen (1955). Civil Aircraft Accident. *Report of the Court of Inquiry into the Accidents to Comet G-ALYP on 10th January, 1954 and Comet G-ALYY on 8th April 1954*, London, UK: Ministry of Transport and Civil Aviation, pp. 1–48.

[9] Oehlke, A. (2005). The English example: the introduction of modern cotton mills and textile technology from Lancashire. In: *Cotton Mills for the Continent*, 23–26. Dortmund, Germany: Westfälisches Industriemuseum (in German).

[10] De Twentsche Courant Tubantia, Enschede, The Netherlands, September 14, 2013, p. 15 (in Dutch).

[11] van den Berg, J. (2012). Fireproof storage of cocoa with nitrogen. *Bulk 20* (4): 34–36. (in Dutch).

[12] van den Berg, J. (2012). Fireproof cocoa warehouses for Cargill. *Technisch Weekblad* 43 (18): 5. (in Dutch).

[13] Wiley Online Library (1999–2013). Flame Retardants. Ullmann's Encyclopedia of Industrial Chemistry. New York: Wiley Online Library.

[14] Sacher, J. (2011). Exemplary by-blow. *Process* 18: 40–41. (in German).

[15] Department for the maintenance of dikes, roads, bridges, and the navigability of canals in The Netherlands (Rijkswaterstaat) and the Royal Dutch Meteorological Institute (KNMI) (1961). *Report on the Storm Surge of 1953*, Staatsdrukkerij- en Uitgeversbedrijf, The Hague, The Netherlands, pp. 14, 16 (in Dutch).

[16] Ringers, J.A. (1948). In: *50 Years, Memorial Volume on the Occasion of the 50th Anniversary of Queen Wilhelmina's Reign*, Staatsdrukkerij- en Uitgeversbedrijf, The Hague, The Netherlands, p. 388 (in Dutch).

[17] van der Ham, W. (2003). *Master of the Sea, Johan van Veen, Engineer in the Department for the Maintenance of Dikes, Roads, Bridges, and the Navigation of Canals, 1893–1959*, 92. Amsterdam, The Netherlands: Uitgeverij Balans (in Dutch).

[18] Dutch Ministry of Infrastructure and the Environment, Maeslantkering, E-mail, December 11, 2014.

[19] Mulder, K. (2011). Chlorofluorocarbons – drivers of their emergence and substitution. In: *What Is Sustainable Technology? Perceptions, Paradoxes and Possibilities*, 22–38. Sheffield, UK: Greenleaf Publishing.

[20] NRC Handelsblad, Rotterdam, The Netherlands, January 6, 2011, p. 4 (in Dutch).

[21] German Federal Ministry for Food and Agriculture (2012), Press Information No. 100, April 3, 2012 (in German).

[22] Siemerink, B. and Sauer, G. (2012). *The Motorcycle Races at Tubbergen (1946–1984)*, 92. Denekamp, The Netherlands: Boekwinkel Heinink (in Dutch).

[23] de Volkskrant, Amsterdam, The Netherlands, June 4, 2013, p. 11 (in Dutch).

4

SAFETY ASPECTS NEED ATTENTION

4.1 INTRODUCTION

This chapter treats six cases concerning transport in Section 4.2 and four cases regarding the society in general in Section 4.3. Safety should have received more attention in all the 10 cases. A short description of these cases in Sections 4.2 and 4.3 follows now. First, the cases in Section 4.2. In the first case, the environment was an important aspect in the selection of natural gas instead of diesel oil for a bus. However, the bus became a flame-thrower in a fire. In the second case, the desire to avoid paying toll played a key role in the decision to buy light trucks with trailers. However, a gust of wind can blow them off the road. In the third case, attention is paid to a car coolant that neither attacks the ozone layer in the stratosphere nor is a greenhouse gas. However, that car coolant is inflammable. The fourth case in Section 4.2 deals with a railway accident in Germany in 1998. Wheels with a tire were selected for Intercity-Express (ICE) trains to avoid vibrations. There were rubber dampers between the wheels and the tires. A tire broke and it was the root cause of a major accident. Deutsche Bahn went back to solid wheels. It is an example of a worst-case situation. What

Safety in Design, First Edition. C.M. van 't Land.
© 2018 John Wiley & Sons, Inc. Published 2018 by John Wiley & Sons, Inc.

started as an incident became worse and worse due to unfortunate circumstances and ended in a catastrophe. The fifth case of Section 4.2 concerns the burning of a lithium ion battery in a Boeing 787 Dreamliner. Working with lithium ion batteries asks for good process control. The sixth case of Section 4.2 concerns a ferry service on the North Sea Canal between Amsterdam and Velzen-Zuid in The Netherlands. The service was provided by fast boats having a velocity of $60 \, km \, h^{-1}$. Accidents happened and it appeared that the operation was not safe.

The first case in Section 4.3 deals with earthquakes related to the production of natural gas in the northern part of The Netherlands. Natural gas exploitation started in 1963. Earthquakes became more violent in the course of the years. However, the safety of the citizens has not been taken into account till the beginning of 2013.

The second case in Section 4.3 deals with a fierce fire in a plant dealing with inflammable materials. Several rules were neglected by the company. A neighboring company was also destroyed. It is advisable to concentrate risky companies geographically. They can then assist and support each other. The third case pleads for the use of noninflammable insulation materials for buildings. Polystyrene is often used; however, it is inflammable. Finally, rolling shutters are discussed. They are good at keeping burglars out but hinder inmates in leaving a house. It is advised to install a manually operated emergency exit.

4.2 TRANSPORT

4.2.1 Bus on Natural Gas Afire at Wassenaar in The Netherlands in 2012

In the morning of October 29, 2012, a bus on natural gas took fire at Wassenaar in The Netherlands [1]. The fire had started in the motor space at the back of the vehicle. The driver noticed smoke coming out of the left side of the vehicle's back, stopped the bus by the side of the road, and asked the 5 passengers to leave the bus. The fire-brigade was noticed. It took the fire-brigade about 11 to 14 min to get to the bus. The fire-brigade supervised the fire but did not try to extinguish as the fire had advanced substantially. Shortly after their arrival, horizontal flashes having lengths between 15 and 20 m left the left side of the roof of the bus (see Figure 4.1). The flashes were comparable to the flashes leaving a flame-thrower. The flashes lasted approximately 4 min. Neither people nor buildings were affected by the flashes. The bus was completely destroyed by the fire.

Figure 4.1 Flashes from a burning bus on natural gas in The Netherlands.
Source: Courtesy of Dutch Safety Board, The Hague, The Netherlands.

The flashes were caused by the combustion of natural gas leaving fuel cylinders on the top of the roof through pressure relief devices mounted on the fuel cylinders. The pressure relief devices directed the flashes horizontally. The natural gas was stored in 8 cylinders on the roof of the bus. The cylinders were located perpendicularly to the length of the bus. The cylinders had been filled with natural gas up to a pressure of 200 bar. The pressure relief devices are under temperature control and open at a temperature of 110 °C. The reason for this safeguarding method is that bursting of the cylinders should be prevented. It can cause hurtling around cylinder fragments and gas explosions.

Conventionally, buses use diesel fuel. The reason to select natural gas is that it is environmentally a better option. Per kilometer, less carbon dioxide, dust, and NO_x (nitrogen oxides) are emitted. Furthermore, buses on natural gas produce less noise than those on diesel fuel.

The introduction of natural gas as a fuel for vehicles started in 2005. In 2013, in The Netherlands, there were slightly more than 600 buses, approximately 2400 trucks (mainly delivery vans), and about 3100 cars using natural gas.

Approximately 0.4% of the buses in The Netherlands take fire annually. There are slightly more than 5100 buses in that country. It means that there are somewhat more than 18 fires in the Dutch buses annually. The Dutch Safety Board estimates that approximately 6 of these 18 fires develop into a serious fire. A distinction between buses on diesel fuel and buses on natural gas has not been made in the aforementioned figures.

The fire could have had more serious consequences in an urban area or in a tunnel. On changing from diesel fuel to natural gas, the safety aspects should have received more attention than they did. The Dutch Safety Board recommends to study the safety aspects of vehicles on natural gas closely in the light of the accident at Wassenaar in 2012 and to come up with improvements. In the past 5 years, worldwide, more similar accidents with buses on natural gas have occurred.

The Dutch Safety Board remarks that both the American and the European regulations on motor vehicles on hydrogen do contain instructions regarding the orientation of the pressure relief devices.

It is striking that a protection against explosions caused a different danger, i.e. long flashes. The safeguarding against explosions, i.e. the pressure relief devices, functioned well.

4.2.2 Light Trucks with Trailers are Dangerous

A truck with a trailer was blown out of a railway carriage by a gust of wind on the Hindenburgdamm in the north of Germany on September 3, 2009 [2]. The Hindenburgdamm connects the German coast with the isle of Sylt. The truck driver was hurtled into the sea and died on the spot due to a severe head injury. Slightly more than 20 accidents in which trucks with trailers were involved and which were caused by gusts of wind occurred in Germany in the past few years (see Figure 4.2).

Figure 4.2 A light truck with trailer were hit by a gust of wind in Germany. *Source:* Courtesy of Harry Härtel/Haertelpress, Chemnitz, Germany.

The trucks with trailers concerned had a light construction. Combinations not exceeding a mass of 12 tons do not have to pay toll in Germany. The obligation for combinations to pay toll in Germany was introduced in 2005. As from that year, light combinations were introduced in that country, and, at the present time, it is estimated that several thousands of light trucks with trailers are in use.

A typical conventional combination has a mass of 15 tons and can load 25 tons. Its length is up to 18.75 m and its height is up to 4 m. A typical light combination has a mass of 7 tons and can load 5 tons. It has approximately the same dimensions as a conventional combination. Whereas the mass of the light trailer itself is less than 2 tons, the side offers an area up to 30 m^2 to the wind. Thus, the light combination, and especially the trailer, is sensitive to the wind. Being able to withstand gusts of wind is not a criterion for approval of combinations of a truck with a trailer in Europe of today.

Light combinations can transport about one-fifth of the mass transported by conventional combinations. Thus, they are suitable for, e.g. the transport of insulation materials and empty plastic containers.

Light combinations are cheaper to buy and operate than conventional trucks with trailers. As for the fuel, conventional combinations consume typically 30 l of diesel per 100 km and combinations having a light construction consume maximum 20 l per 100 km.

A German transport company operating light combinations states that they are aware of the wind sensitivity. They cope with this situation by ordering the drivers to stop and bring their light combinations to a safe location when the wind is strong. The decision to stop is taken centrally and communicated to the drivers. There is a differentiation between loaded and empty combinations. The decision to stop is taken on the basis of the predictions of the Internet program Windfinder.

This section started with the description of an accident on the Hindenburgdamm in Germany. A German court passed a sentence on the railway company because of its negligence. It is a German regulation that the combination should have been fastened for the transport. Such fastening had not been executed. The train driver asked the truck driver whether the combination was fully loaded. The answer was affirmative and the train driver then decided not to fasten the combination. The verdict of the court appears to be correct. The fastening has to be applied strictly and is to be safeguarded. Trucks having a light construction are more difficult to protect against the hazard of windy roads than against the hazard of a gust of wind during rail transport.

The accidents cannot be explained by an increase of the average wind velocity in Germany.

4.2.3   Car Refrigerants

In 2012, the German car manufacturer Daimler carried out tests concerning the inflammability of a new car refrigerant at Sindelfingen in Germany [3, 4]. The name of the new refrigerant is R1234yf. They simulated a leakage of the refrigerant from the lower parts of the air conditioning unit in the motor space while the motor was hot. At these tests, the cars had not collided with an object and were thus undamaged. The cars took fire. Daimler thereupon questioned the correctness of the decision to introduce the new refrigerant.

Back in the 1980s, R12 was a widely used car refrigerant. As it is a CFC (see Section 3.4.2), it attacks the ozone layer in the stratosphere on escaping and so was banned. In the 1990s, it was replaced by R134a, an HFC. The latter material does not attack the ozone layer, is nontoxic and noninflammable. However, it is a greenhouse gas (see Section 3.4.2) with a 100-year global warming potential (GWP) 1430 times greater than carbon dioxide. Efforts were put to find an alternative for R134a that would neither attack the ozone layer nor be a greenhouse gas and would be nontoxic.

In the first years of this century, Honeywell and DuPont proposed an alternative: R1234yf. It does not attack the ozone layer, has a 100-year GWP four times the GWP of carbon dioxide, and is nontoxic. However, unlike R12 and R134a, it is mildly inflammable. In addition, when R1234yf burns, hydrogen fluoride is formed. The latter compound is an etching material and dangerous for people, e.g. rescue forces.

All the three abovementioned materials have the right thermodynamic properties; their boiling points at atmospheric pressure are in the range of -25 to $-30\,°C$. Thus, car air conditioning units that are suitable for R134a are, in principle, also suitable for R1234yf.

Although being mildly inflammable, R1234yf was approved by the European Union. This approval was questioned by Daimler in 2013. The European Commission's Joint Research Centre concluded in March 2014 that there is no evidence of a serious risk in the use of R1234yf in mobile air conditioning (MAC) systems.

There is a potential car refrigerant that satisfies almost all requirements – carbon dioxide. However, carbon dioxide is a greenhouse gas (see Section 3.4.2). Moreover, carbon dioxide cannot be used in conventional car air conditioning units. An important reason is that the maximum pressure in air conditioning units operating with carbon dioxide is 100 bar. The maximum pressure in air conditioning units operating with the other two refrigerants (R134a and R1234yf) is 10 bar. Air conditioning units operating with carbon dioxide will have to be designed.

4.2.4 The Eschede Train Accident in Germany in 1998

The Event At 10.59h on June 3, 1998, the Eschede train accident occurred near the village of Eschede in Germany [5, 6]. Eschede is near the town of Celle. In the accident, 101 people died and approximately 100 were injured. The train concerned was an Intercity-Express (ICE) running from Munich in the south of Germany to Hamburg in the north. The train consisted of a front power car, cars 1, 2, 3, 4, 5, 6, 7, 9, 10, 11, 12, and 14, and a rear power car.

Six kilometers south of Eschede the steel tire on a right wheel of the first car of the train broke and peeled away from the wheel. The train had a speed of 200 km h^{-1} at that time. The broken tire punctured the floor of the first car; however, an important part remained under the floor in the vicinity of the rotating wheels. The train then covered a distance of approximately 5.5 km. Subsequently, the train passed over the first of two track switches approximately 200 m before the location where a traffic bridge crossed the railway. The broken tire carried the guide rail of the first switch away, which also punctured the floor of the first car. This event caused the derailment of the damaged wheel of the first car. The left wheel, opposite to the damaged wheel, also derailed and hit the points lever of the second switch, changing its setting. This event caused the derailment of the first four cars. The front power car and the first two cars cleared the bridge. The third car derailed violently and hit the piers of the bridge, which thereupon collapsed. The third car cleared the bridge. The fourth car also cleared the bridge and rolled intact into the embankment immediately behind it. The front part of the fifth car cleared the bridge as well; however, the rear part was crushed by the falling bridge. The sixth car, the restaurant coach, was also crushed by the falling bridge. The remaining cars all derailed and collected against the rubble in a zigzag pattern. The rear power car came to a standstill at an angle of about 90° with the remaining cars.

Wheel Design The ICE-train concerned was equipped with Dual Bloc wheels. These wheels consisted of a steel wheel body surrounded by a 20-mm-thick rubber damper followed by a metal tire (see Figure 4.3). The Dual Bloc wheels had replaced Monobloc wheels, which were single-cast wheel sets. The reason for this replacement was that the latter wheels caused, at cruising speed, vibrations. These vibrations caused loss of comfort and it was feared that parts of the train would be damaged. The replacement, which was approved by the management of Deutsche Bahn in 1992, resolved the issue of vibration. A design resembling the design chosen by Deutsche Bahn had been a success in, for instance, trams

Figure 4.3   A Dual Bloc wheel.

at Hannover in Germany. However, üstra, the company that operates Hannover's trams, discovered fatigue cracks in Dual Bloc wheels in July 1997. It began changing wheels before fatigue cracks could develop. The new wheels were also Dual Bloc wheels. üstra reported its findings as a warning to all other users of wheels built with similar design, including Deutsche Bahn, in autumn 1997. According to üstra, Deutsche Bahn replied by stating that they had not noticed problems in their trains [7].

Additional Remarks   It strikes that the seriousness of the accident was caused by several aspects. "The worst-case situation" started with the collapse of the tire on a wheel of the first car. A breakage of a tire on a wheel of, for instance, the last car would probably have had less serious consequences. Furthermore, the train rode on a track not specifically built for the ICE-train. The track was also used by other train types. Hence, the train had a speed of $200 \, \mathrm{km} \, \mathrm{h}^{-1}$, which was the maximum speed for the ICE-train on this specific track. The presence of switches contributed to the seriousness of the accident. The presence of a traffic bridge crossing the railway immediately behind the switches is a further aspect. The bridge had two piers that were hit by the third car which caused the collapse of the bridge. If this bridge would not have had two piers but would have been supported on the two embankments, the accident would probably have been less serious.

Concluding Remarks The wheels of a train are, safetywise, critical parts. It is not possible to mitigate the effects resulting from breakage of a wheel tire. The proper functioning of the wheels was safeguarded by regular inspections. This method failed as the weakness of the tire was not detected during an inspection at Munich the night before the accident happened. Deutsche Bahn replaced the Dual Bloc wheels on all ICE-trains concerned by Monobloc wheels in the weeks following the accident.

4.2.5 Burning Battery in Boeing 787 Dreamliner in 2013

The Events The first commercial flight of a Boeing 787 Dreamliner airplane was on October 26, 2011. A battery of a Boeing 787 Dreamliner became overheated and a fire started in a battery compartment of the empty airplane at Boston's Logan International Airport on January 7, 2013 [8]. The airplane was operated by Japan Airlines (JAL).

On January 9, 2013, United Airlines reported a problem with the wiring in one of its six 787s in the same area as the area of the battery fire on JAL's airliner on January 7, 2013.

An All Nippon Airways (ANA) 787 made an emergency landing at Takamatsu Airport on Shikoku Island in Japan on January 16, 2013 after the flight crew received a computer warning that that there was smoke inside one of the electrical compartments. According to ANA, there was an error message in the cockpit citing a battery malfunction.

JAL and ANA, two major Japanese airlines, announced that they were voluntarily grounding or suspending flights for their 787s on January 16, 2013. Prior to that date, 24 Dreamliners had been delivered to these two airlines. The Federal Aviation Authority (FAA) in the United States ordered United Airlines, an American airline, to ground their six 787s as well. Fifty Dreamliners in all had been sold by Boeing on January 16, 2013. Other countries ordered their airlines to also ground their 787s. The grounding by the FAA was the first time FAA grounded an airplane type since 1979.

On April 19, 2013, the FAA gave United Airlines permission to use their 787s again. On that date, changes had been made to the battery systems. In April 2013, the Japanese authorities also decided to allow Japanese Dreamliners to return to service.

On January 14, 2014, JAL reported that a maintenance crew at Narita Airport at Tokyo in Japan discovered smoke coming from the main battery of one of its Boeing 787 jets, 2 h before the plane was due to fly to Bangkok. Maintenance workers found smoke and unidentified liquid coming from the main battery, and alarms in the cockpit indicated faults with the power pack and its charger.

Battery Design   The Dreamliner contains two sets of eight batteries, each set weighing 30 kg. One set serves to start an auxiliary power unit, a small generator raising the electric power to start the jet engines. The other set serves as backup for systems on board the airplane. Boeing has made the airplane Boeing 787 Dreamliner as light as possible. Boeing claims that this leads to, compared to competing airplanes, a reduction of the fuel consumption of 20%. The selection of lithium ion batteries comes within the compass of this effort. However, lithium ion batteries are inflammable. This is caused by the presence of an organic solvent in the cells. When the temperature within the battery becomes too high, the organic solvent decomposes and the developed gases can take fire.

On discharging a battery, chemical energy is converted into electrical energy. On charging a battery, the reverse process takes place. Both processes are always accompanied by some heat development. If a battery is mistreated, the heat development can be substantial.

Within the category of lithium ion batteries, Boeing selected lithium ion batteries having a positive electrode made of lithium cobalt dioxide. This battery type has an energy density of 150 Wh kg^{-1} battery weight, which is a relatively high figure. However, the risks associated with this specific battery type are greater than those with, for instance, lithium ion batteries having a positive electrode made of lithium iron phosphate (LFP batteries). The energy density of the latter type of batteries is approximately two thirds of the former type of battery. The electrode made of lithium iron phosphate is inherently safer than the electrode made of lithium cobalt dioxide due to its greater structural stability and because oxygen in its structure is relatively strongly bound. These two aspects make the LFP battery more resistant to misuse (in particular on charging it) than the battery selected by Boeing.

Each set of batteries contains eight batteries next to each other. On January 7, 2013, the fire at Logan airport at Boston was caused by the overheating of one of the central batteries of one of the two sets of batteries. Subsequently, the organic solvent in the battery concerned decomposed and the gases developed thereby took fire. The other batteries of the set of batteries concerned were then overheated as well. The set of eight batteries was completely destroyed (see Figure 4.4).

Figure 4.4 NTSB Materials Engineer Matt Fox examines the casing of the battery involved in the JAL Boeing 787 fire incident at Boston. *Source:* Courtesy of National Transport and Safety Board, Washington, DC, USA.

Overheating of this type of battery can occur if the battery is over-charged. Overheating of this type of battery can also occur if the battery is ran down too far. Process control has to take care that neither overcharging nor running down too far can occur. A further aspect is that installing the batteries next to each other means that a central battery cannot get rid of heat in case overheating of the battery contents occurs. Heat is transferred from one battery to the next.

Boeing outfarmed the manufacture of the batteries to the Japanese company GS Yuasa. Neither Boeing nor GS Yuasa has informed the public of the cause of the fire on January 7, 2013, possibly because the cause of the overheating is unknown.

Concluding Remarks Lithium ion batteries having lithium cobalt dioxide as positive electrode are used for many purposes. Their use can be considered safe if the processes within the battery are controlled properly. It is possible that the process control of the batteries in the Dreamliner was not adequate in January 2013. It is also possible that Boeing had the process control improved after the fire and incidents described previously. Boeing communicated that the following measures were implemented:

- The improvement of electrical wires and connectors in the battery compartments

- The installation of insulating layers between the batteries of a set of eight batteries
- The installation of an insulating layer around each battery
- The installation of a stainless steel hull to enclose each set of eight batteries and
- The installation of a pressure relief valve on the stainless steel hulls relieving to the atmosphere outside the airplane. Thus, in case of decomposition, the access of oxygen is prevented and a fire cannot develop.

I started to study chemical engineering at the University of Twente in The Netherlands in 1964. There were, at that time, three faculties, namely, chemical engineering, electrical engineering, and mechanical engineering. A student could complete the first year at the university without making a choice for a faculty. Thus, students acquired knowledge of all three fields. When I worked for Akzo Nobel, I noticed that it has, for engineers, advantages to also have knowledge of other fields than their own field. It is possible that having more knowledge of electrochemistry than they had would have paid off for the aeronautical, electrical, and mechanical engineers at Boeing. It could then have been realized that proper process control is of paramount importance for the safe operation of lithium ion batteries having a positive electrode made of lithium cobalt dioxide.

4.2.6 Ferry Service on the North Sea Canal in The Netherlands

Events A ferry service existed on the North Sea Canal in The Netherlands between Amsterdam and Velzen-Zuid in the period April 1998 to January 1, 2014 [9] (see Figure 4.5). The length of the reach was approximately 20 km and the average width of the North Sea Canal is 280 m. A number of harbors are connected to the canal between Amsterdam and Velzen-Zuid. The boats used were so-called hydrofoils. A hydrofoil's hull is lifted out of the water as speed is gained, whereby drag is decreased and speeds in the range 50–65 km h^{-1} are made possible. The initial service speed of the hydrofoils was 65 km h^{-1}. Later on, the service speed was decreased to 60 km h^{-1} and that speed was maintained till January 1, 2012. The maximum speed of all other boats on the greater part of the North Sea Canal is 18 km h^{-1}. Five accidents in which the boats were involved occurred in the period 2003 up till and including 2008. Hydrofoils hit a bank in three accidents, in one

Figure 4.5   A hydrofoil. *Source:* Courtesy of Connexxion Holding NV, Hilversum, The Netherlands.

accident a hydrofoil hit a second boat, and a second boat hit a hydrofoil in one instance. The most serious accident, in which 21 people were injured, occurred on October 18, 2003. A bank was hit in this accident. A further accident, in which one person was seriously injured and several people were injured, occurred on October 8, 2007. A hydrofoil was hit by a different ship in this accident.

The Dutch Safety Board investigated these accidents and concluded that the transport by these fast boats was not adequately safeguarded. A general rule is that fast boats have to give way to slower boats. The Board stated in 2009 that giving way to other boats, when a hydrofoil has a speed of $60\,km\,h^{-1}$, is not fully possible. The skipper must, in too short a time, and at too large a distance from the object or ship to be avoided, start a corrective action to evade the object or ship. Speed reduction is not the solution as the stability and the maneuverability of a hydrofoil decrease when the speed is reduced to a value in the range 20–$45\,km\,h^{-1}$. Evading an object or ship then becomes quite problematic.

The service speed of the hydrofoils was reduced to $50\,km\,h^{-1}$ on January 1, 2012, to prevent further accidents. The service was stopped on December 31, 2013. The speed reduction to $50\,km\,h^{-1}$ caused a decrease of the number of trips per day. This decrease led to a decrease in the number of passengers per day and that hurt the economy of the service.

Additional Facts   Connexxion, the operator of hydrofoils, received an exemption from the maximum speed on the North Sea Canal. The maximum speed is $18 \, km \, h^{-1}$ on the greater part of the canal.

Ships having a speed higher than $40 \, km \, h^{-1}$ always have to give way to other ships on the North Sea Canal. The background of this rule is that it is not easy for ships having a relatively low speed to give way to ships having a speed higher than $40 \, km \, h^{-1}$. The Dutch Safety Board concluded that it is not fully possible for hydrofoils having a speed of $60 \, km \, h^{-1}$ to always give way to slower boats. The Board illustrates this with figures. A fast ferry having the service speed of $60 \, km \, h^{-1}$ has advanced $139 \, m$ when a port correction of $20 \, m$ has been made. It takes the boat $8 \, s$ to advance $139 \, m$ at the service speed. A hydrofoil having the service speed of $60 \, km \, h^{-1}$ has advanced a distance between 190 and $280 \, m$ when it stops. For cars having a speed of $60 \, km \, h^{-1}$, this distance is $45 \, m$ on a wet road.

Concluding Remark   Economy and trip time received attention before the introduction of hydrofoils on the North Sea Canal. The trip time was $27 \, min$ when the hydrofoil's speed was $60 \, km \, h^{-1}$. The Dutch Safety Board concludes that safety should have received more attention before the introduction than it did.

4.3   SOCIETY

4.3.1   Earthquakes Related to the Production of Natural Gas in the Northern Part of The Netherlands

Events   The production of natural gas in the northern part of The Netherlands has caused earthquakes in that part [10]. The first relatively important earthquake that was thought to be related to the extraction of natural gas occurred at Assen in the province Drente in 1986. Both provinces Drente and Groningen are in the northern part of The Netherlands. It had a magnitude of 2.7 on Richter's Scale. The first relatively important earthquake in the province Groningen that was thought to be related to the production of natural gas occurred at Middelstum in 1991. It had a magnitude of 2.4 on Richter's Scale. The annual number of earthquakes above the Groningen reservoir increased between 1991 and 2012. However, the annual average magnitude of these earthquakes did not increase in that period [11]. On August 16, 2012, an earthquake having a magnitude of 3.6 on Richter's scale occurred at Huizinge in the province Groningen. That earthquake is the most serious one experienced until now. Neither people nor animals were harmed. Buildings did not collapse; however, a lot of damage was done to many buildings. The damage was greater compared to that of earlier earthquakes. People panicked during the Huizinge earthquake and fled into the streets.

The occurrence of the most serious earthquake until now at Huizinge in 2012 and the fact that the annual average magnitude of the earthquakes did not increase between 1991 and 2012 can be understood as follows. 584 earthquakes of a magnitude greater than 1 on Richter's Scale have occurred until now [11]. These earthquakes have various magnitudes, e.g. 1.5, 2.2, and 1.8. A probability distribution of magnitudes can be made. The Huizinge earthquake had a magnitude of 3.6. An earthquake having a magnitude of 3.6 could have happened in, e.g. 1995. By the same token, an earthquake having a magnitude of, e.g. 4.5 can happen in 2020. It is a matter of probability.

There were 187 damage reports in 2011, and NAM, the company exploiting the natural gas reservoir, paid € 560 359 to compensate for the damage. There were 13 384 damage reports in 2014. The amount of money paid to compensate for the damage is not yet known.

The Production of Natural Gas in the Northern Part of The Netherlands   In 1959, a large reservoir of natural gas was discovered in the province Groningen in The Netherlands. It is, with an initial capacity of 2800 billion nm^3, Number 9 on the list of most important natural gas reservoirs worldwide. The largest reservoir of natural gas on the earth is the South Pars reservoir in Iran and Qatar. It had an initial capacity of 10 000– 15 000 billion nm^3. The reservoir in Groningen is the only large natural gas reservoir worldwide under a relatively densely populated area. The area over the reservoir is approximately 900 km^2. Figure 4.6 gives an impression of the size of the gas reservoir. It is a map showing the expected subsidence in cm due to the gas production in Groningen in 2070. The exploitation of natural gas in Groningen started in 1963. The reservoir is present at a depth of approximately 3 km. The initial pressure of the gas in the reservoir was approximately 320 bar. The pressure in the spring of 2015 was approximately 100 bar [9]. Approximately three quarters of the initial amount of natural gas have been extracted. Almost 30 billion nm^3 of natural gas were produced in 1990, whereas 53.2 billion nm^3 of natural gas were extracted in 2013. The latter amount was the largest annual production from the reservoir since the production started.

Earthquakes   Two categories of earthquakes can be distinguished. First, natural or tectonic earthquakes. A great majority of earthquakes occurring

Figure 4.6 Subsidence in cm expected in Groningen in 2070. *Source:* Courtesy of Nederlandse Aardolie Maatschappij B.V., Assen, The Netherlands.

on the earth are natural earthquakes. The heaviest natural earthquake recorded until now was the one that measured a magnitude of 9.5 on Richter's Scale (Chili 1960). Natural earthquakes having a magnitude greater than 8 occur on the earth on average once a year. Induced earthquakes form the second category. They are caused by human activities deep in the earth. The earthquakes experienced in the northern part of The Netherlands are induced earthquakes.

The magnitude is one parameter to characterize the effect of an earthquake. The magnitude is expressed by a number on Richter's Scale ranging from 0 to 10. It is a logarithmic scale; thus, an earthquake that measures 4 on this scale has a magnitude 100 times greater than an earthquake of magnitude 2. Furthermore, the effect of earthquakes is characterized by two further parameters, i.e. the energy and intensity. The energy of an earthquake increases strongly with the number on Richter's Scale. An earthquake having a magnitude of 4 on Richter's Scale and starting at a depth of, e.g. 3 km has a greater intensity at the earth's surface than an earthquake having the same magnitude and starting at, e.g. a depth of 10 km. The Groningen reservoir, having a depth of approximately 3 km, is relatively close to the earth's surface.

Knowledge of Earthquakes in Groningen Not much knowledge has been acquired worldwide concerning earthquakes caused by the exploitation of natural gas reservoirs. One reason is that large natural gas reservoirs present in the earth's soil are under relatively uninhabited areas. When the production of natural gas in Groningen started in 1963, it was thought that the exploitation would lead to subsidence of the soil only and that earthquakes would not occur. However, earthquakes related to the production of natural gas did occur. Natural earthquakes do not occur in this part of The Netherlands. The first relatively significant earthquakes were experienced in 1986. A commission of the Dutch government issued a report in 1993 in which it was stated that the maximum magnitude of earthquakes in the northern part of The Netherlands due to the production of natural gas would be 3.3 on Richter's Scale. However, an earthquake at Roswinkel in the province Drente in 1997 had a magnitude of 3.4. NAM, the company exploiting the natural gas reservoir, received more than 200 damage reports related to that earthquake. The Royal Dutch Meteorological Institute issued a report in 1998 in which it was stated that the maximum magnitude of earthquakes in the northern part of The Netherlands would be 3.8 on Richter's Scale. The incentive to issue this report was the earthquake at Roswinkel in 1997. The Royal Dutch Meteorological Institute issued a further report in 2004 stating that the maximum magnitude of earthquakes in the northern part of The Netherlands would be 3.9 on Richter's Scale. An earthquake having a magnitude of 3.5 on Richter's Scale occurred at Westeremden/ Middelstum in the province Groningen in 2006. An earthquake of magnitude 3.6 occurred at Huizinge in the province Groningen in 2012, which was mentioned earlier.

The recapitulated sequence of magnitudes of earthquakes that actually occurred and the sequence of statements concerning the maximum magnitudes to be expected prove that the knowledge of earthquakes due to the production of natural gas in the northern part of The Netherlands is inadequate. A program to acquire the necessary knowledge has been started in January 2013. The view is held at present that a maximum magnitude of earthquakes to be expected exists. However, it can be greater than 3.9 on Richter's Scale.

Summary of the Earthquakes in the Northern Part of The Netherlands The production started in 1963.

Magnitudes on Richter's Scale of major earthquakes between 1986 and 2012.

1986	Assen	2.7
1991	Middelstum	2.4
1997	Roswinkel	3.4
2006	Westeremden/Middelstum	3.5
2012	Huizinge	3.6

584 earthquakes of a magnitude greater than 1 on Richter's Scale occurred between 1991 and 2015.

Predictions of maximum magnitudes on Richter's Scale of earthquakes between 1963 and 2015

1963	subsidence only
1993	3.3
1998	3.8
2004	3.9
2015	>3.9

Complaints received

2011	187
2014	13 384

Concluding Remarks   NAM, the company exploiting the natural gas reservoir in the northern part of The Netherlands, and the Dutch government, considered, till the beginning of 2013, the risk of the production of natural gas in the northern part of The Netherlands the occurrence of small damages. It was considered that such small damages could be simply compensated. The Huizinge earthquake in 2012 showed that the damage caused by earthquakes could be greater and that the safety of the citizens of Groningen is at stake. The safety of the citizens of Groningen was not taken into account till the beginning of 2013.

4.3.2   Fire at Chemie-Pack at Moerdijk in The Netherlands in 2011

Event   A fire started at the site of a company called Chemie-Pack at Moerdijk in The Netherlands at approximately 14.20 h on January 5, 2011 [12]. The fire spread rapidly due to the ignition of inflammable materials.

The site was completely destroyed by the fire. The adjacent site of a company called Wärtsilä, a Swiss company manufacturing diesel engines for ships, was also completely destroyed by the fire.

Chemie-Pack   Chemie-Pack was an independent company, employing approximately 50 people, that processed chemicals and stored them. The processing of chemicals was physical and encompassed blending and packaging. Many of the materials processed by Chemie-Pack were inflammable.

Chemie-Pack was a BRZO-company in The Netherlands. BRZO stands for Decision Risks Heavy Accidents. A BRZO-company refers to a company that processes or stores large amounts of dangerous materials. There are more than 400 BRZO-companies in The Netherlands.

Cause of the Fire   An air-driven membrane pump was used in the open air to transfer a resin in the afternoon of January 5, 2011. A tray containing xylene was under the membrane pump. The xylene originated from cleaning activities that had been carried out prior to the transfer of the resin. The ambient temperature was 3–4 °C. The activities to transfer the resin started at approximately 13.00 h. Shortly after the start, problems were encountered with the pump. The pump did not transfer resin. The air from the air-driven pump passed through a muffler, in which icing had occurred. That was the cause of the interruption of the transfer. The first step taken by the operator was to increase the pressure of the air driving the pump to 7 bar. This pressure was adjustable in the range 2–7 bar, a low pressure corresponded with a low output and vice versa. The second step was heating the muffler with a flame. The flame came from a provision to apply plastic wraps around packagings by shrinking. The second step was successful and the transfer resumed. However, shortly after 14.00 h, it was again noticed that the pump did not transfer the resin. It is unknown whether this was due to icing of the muffler or due to an obstruction in the resin discharge line. The operator then heated not only the muffler but also the pump's body with the flame. This caused the ignition of xylene in the tray under the membrane pump (see Figure 4.7). The flash point of xylene is approximately 20 °C. The membrane pump was not stopped! During attempts of Chemie-Pack employees to extinguish the fire, a flash was noticed. A rupture in the discharge line close to the pump had occurred. Resin was pumped through an opening and was ignited by the burning xylene. The cause of the rupture was probably the combination of high air pressure and pump heating. These two aspects caused a high pressure in the discharge line because the line was plugged. The plugging was probably

Figure 4.7 The fire started at the membrane pump. *Source:* Courtesy of Police Zeeland – West-Brabant, Tilburg, The Netherlands.

caused by the cooling of the resin in the discharge line, leading to a high viscosity of the resin. The membrane pump discharged the resin through the opening not only in the tray, but also outside the tray.

Fire-fighting Activities The site of Chemie-Pack did not have an own fire-brigade. The first action was taken by Chemie-Pack personnel trying to use a powder extinguisher. However, the powder extinguisher failed to function. Further attempts with available extinguishers failed as well, mainly because it was not possible to extinguish the burning resin. Next, an employee tried to extinguish the fire with a water jet. This attempt made things worse as burning xylene and resin were transferred to IBC-containers containing inflammable materials in the vicinity of the membrane pump. Thereupon, the IBCs took fire.

The company had raised an alarm at 14.26 h. The first car of the fire-brigade arrived at 14.35 h and further seven fire-brigade cars arrived at 14.43 h. The commanding officer gave the order to contain and control the fire. Attempts to extinguish the fire were not made. The fire-brigade declared the fire under control shortly after 00.00 h on January 6, 2011.

Additional Remarks Chemie-Pack's permission to work mentioned that it was not allowed to use flames. The permission to work also mentioned that it was not allowed to store inflammable materials in the open space of the site. The membrane pump was installed in the open space of the site. Approximately 120 IBC-containers with inflammable materials had been stored in the open space.

Icing of the muffler of the membrane pump had occurred earlier. Provisions such as additional drying of the compressed air or warming the compressed air to avoid icing problems had not been installed.

Laws, rules, instructions, agreements, permissions, institutions, and inspections exist in The Netherlands. Nevertheless, this fire occurred because the regulations were neither observed nor enforced. This section is a recapitulation of what happened in this plant.

Approximately 120 IBC-containers were present in the open space of the site. They contained inflammable materials. IBC stands for intermediate bulk container. These containers were made out of high-density polyethylene (HDPE) and in the form of a cube. They were placed inside a gauze-like metal frame for protection. HDPE softens at 70 °C and melts at temperatures in the range 105–130 °C. Thus, plastic IBCs are very vulnerable in case of fire. Metal IBCs exist as well.

Concluding Remarks VNCI, the Association of the Dutch Chemical Industry, has made the suggestion to, in industrial areas, combine chemical plants. This would enable companies to support and assist each other [13].

A further advantage is that companies not dealing with chemicals will not be affected.

4.3.3 Inflammable Building Insulation Material

Event On May 29, 2012, a fire ignited in a new apartment building at Frankfurt am Main in Germany [14]. The apartment building had not yet been put into use. The building had six floors. The fire was fierce, and it took 80 men of the fire-brigade to extinguish it. The cause of the sharpness of the fire was the presence of polystyrene sheets between the walls of the building for insulation purposes. These sheets accelerated the fire. Mineral insulation materials had been alternated with polystyrene sheets along the height of the building to prevent spreading of the fire. However, at this fire, the presence of these noninflammable insulation materials proved useless as the fire passed them readily. It was difficult for the fire-brigade to extinguish the fire because the polystyrene sheets were located between two walls.

Polystyrene Polystyrene is, like all organic materials, inflammable. The polymer softens on heating, and, at approximately 100 °C, the glass transition point is reached. On further heating, the material liquefies. This fact adds to the fierceness and spreading of a fire as the burning liquid flows.

Concluding Remarks   Polyurethane, polyisocyanurate, and mineral insulation materials are alternatives for polystyrene. The first two materials are inflammable as well; however, they do not liquefy on burning. Mineral insulation materials are noninflammable and, thus, a good choice possibly for many applications.

4.3.4   Rolling Shutters

The Event   A fire ignited in a house at Cuijk in The Netherlands on June 20, 2013. Three women, a mother and two daughters, died in the fire. The house was equipped with rolling shutters. The three women did not manage to open the rolling shutters. Neighbors could not reach them because of the rolling shutters.

Rolling Shutters   The rolling shutters in the house concerned could be operated electrically. However, it is not known whether the rolling shutters could be activated by the women. It is possible that the fire had damaged the electrical system. Rolling shutters are adequate to hinder burglars; however, they present a serious risk in case of fire.

Concluding Remarks   Rolling shutters can be operated mechanically, electrically, or both. However, opening the rolling shutters takes time and, in case of fire, time is precious.

It is possible to install a manually operated emergency exit in a rolling shutter. A rope is pulled and the emergency exit door swings open without delay. This safeguarding method is a passive protection. A company like Deelen at Wageningen in The Netherlands can install such a provision, the Innosafe rolling shutter.

To cope with the situation that the electrical system of the house fails in case of fire, it is possible to install a local battery. However, this protection method is an active protection.

It is possible to install a smoke detector. The smoke detector automatically opens rolling shutters if smoke is detected. However, this safeguarding method is also an active protection.

Rolling shutters present few problems to the fire-brigade. They have chainsaws to open rolling shutters. It takes the fire-brigade only a couple of seconds. However, the existence of the fire-brigade is a procedural safety method. The fire-brigade may come in too late.

REFERENCES

[1] Dutch Safety Board, The Hague, The Netherlands: *Fire in a Bus on Natural Gas*, (2013) p. 1–67 (in Dutch).

[2] Wüst, C. (2013). Fear of flying in a truck with trailer. *Der Spiegel* 67 (13): 128–129. (in German).

[3] Wüst, C. (2013). Rebellion of the car manufacturers. *Der Spiegel* 67 (26): 138–140. (in German).

[4] Wüst, C. (2013). Inflammable refrigerant. *Der Spiegel* 67 (35): 139. (in German).

[5] Preuss, E. (1984). *Eschede, 10 Hours 59 – Account of a Railway-Catastrophe*, 1–127. Munich, Germany: GeraMond Verlag GmbH (in German).

[6] Gless, F. and Metzner, W. (2001). The train of death – anatomy of a catastrophe. stern 54 (34): 23–36. (in German).

[7] Wikipedia (2015). Eschede train disaster.

[8] Wikipedia (2015). Boeing 787 Dreamliner battery problems.

[9] Dutch Safety Board, The Hague, The Netherlands, *The Safety of Public Transport with Hydrofoils on the North Sea Canal and the IJ*, (2009), pp. 1–108 (in Dutch).

[10] Dutch Safety Board, The Hague, The Netherlands, *Risks of Earthquakes in Groningen*, (2015), pp. 1–141 (in Dutch).

[11] Biesboer, F. (2015). Majority of earthquakes still to come. *De Ingenieur* 127 (5): 7. (in Dutch).

[12] Dutch Safety Board, The Hague, The Netherlands, *Fire at Chemie-Pack at Moerdijk*, (2012), pp. 1–194 (in Dutch).

[13] te Roller, E. (2011). Fire under control. *Chemie Magazine* 52 (9): 22–24. (in Dutch).

[14] Bartsch, M., Loekx, M., and Ludwig, U. (2012). Seas on fire. *Der Spiegel* 66 (26): 44–46. (in German).

5

MAKE ACCIDENTS AND INCIDENTS VIRTUALLY IMPOSSIBLE

5.1 INTRODUCTION

It is frequently possible to make designs that make accidents and incidents virtually impossible. Those designs can replace designs that are vulnerable and cannot stand up against human failure and equipment failure. A good example of this approach can be found in Section 5.2.8. It deals with the presence of toxic components in the air in cabins of airplanes. It is described in this section that a new design is safetywise superior to the conventional design. Several cases with regard to transport are discussed in Section 5.2, whereas cases regarding the society in general are treated in Section 5.3.

5.2 TRANSPORT

5.2.1 Bus Accident near Barcelona in 2009

Event In the evening of July 30, 2009, a bus having two decks returned from a daytrip to Barcelona in Spain and rode northwestward on highway C-32 [1, 2]. There were 65 passengers on board, of which 64 having the

Safety in Design, First Edition. C.M. van 't Land.
© 2018 John Wiley & Sons, Inc. Published 2018 by John Wiley & Sons, Inc.

Dutch nationality and 1 having the German nationality. The passengers were tourists. The driver took the exit Sant Pol de Mar at 23.10 h. The bus could not follow the exit in a bend, rode through terrain, fell over on its left side, and crossed a road leading to the highway. It then broke through a crash barrier. The bus also collided with at least one other car. Six people died in the accident, whereas almost 40 people were injured. The driver was severely injured.

Background of the Accident   The maximum speed of this bus on a Spanish highway was $100 \, km \, h^{-1}$. The prescribed maximum velocity for this exit was $40 \, km \, h^{-1}$. According to the report of the Catalonian police, the driver drove too fast on the exit. It is probable that the driver took the wrong exit as he should have taken the exit Calella, which is 5 km down the track from the exit Sant Pol de Mar. The bend of that exit is less sharp than that of the exit Sant Pol de Mar. The driver overtook several buses shortly before leaving the highway via the exit Sant Pol de Mar.

Putting on a safety belt was obligatory on board this bus. However, many passengers had not put on their safety belt. It is reported that those passengers who had put on their safety belt were practically free from injuries.

Additional Remarks   The driver drove too fast on the exit. Many passengers had not put on their safety belt. A further aspect is that the bend of the exit was sharp. The prescribed maximum velocity on the exit was $40 \, km \, h^{-1}$. The prescribed maximum velocity on such an exit in, for instance, The Netherlands is $50 \, km \, h^{-1}$. The design of exits should take safety aspects into account. Cars leaving the highway still have high speeds and it takes time to reduce the speed. Sharp bends in exits have to be avoided in particular when icy roads have to be taken into account.

5.2.2   Bus Accident in Hungary in 2003

Event   A German bus was hit by a fast train on a level crossing without crossing barriers at Siófok in Hungary at approximately 09.00 hours on May 8, 2003 [3, 4]. The speed of the train was approximately $100 \, km \, h^{-1}$, and the speed of the bus was very low. The level crossing was equipped with functioning flashing red lights. The train split the bus into two parts. One of the two parts was entrained by the train that came to a standstill about 150 m from the level crossing. The other part was not entrained but took fire and burnt itself out. The locomotive and the first car of the train derailed. A total of 36 German tourists were on board the bus. Thirty-three bus passengers and the bus driver died in the accident, whereas 3 bus passengers were injured. The train driver was injured, whereas the train passengers were unharmed.

Background of the Accident The bus driver possibly tried to stay in touch with buses riding in front of him. He overtook a minibus waiting before the level crossing. Waiting times can be rather long in Hungary. Because of this maneuver, he could only cross the level crossing at a very low speed. He could have noticed both the approaching train and the flashing light.

Two general remarks are made. It is only allowed to cross a level crossing if it is certain that the other side can be reached. It is not allowed to cross a level crossing when flashing lights are present.

Additional Remarks The accident is due to human failure. It would probably not have happened if the level crossing would have been designed with crossing barriers.

5.2.3 Two Train/Truck and Trailer Collisions at Gronau in Germany in 2011 and 2013

Event Number 1 A truck and trailer had loaded 12 metric tonnes of flour from a warehouse at the end of Presterkamp on March 24, 2011 (see Figure 5.1) [5]. The combination then moved backward through Presterkamp and on to the Ochtruper Strasse in the direction of a railroad crossing shortly before 09.20 h. The railroad is a single track one. The driver wanted to enter the Ochtruper Strasse in this way in order to be able to move forward in the opposite direction as a next step. The driver stopped the combination on the railroad crossing. An approaching train hit the trailer subsequently. The train driver and 12 passengers were slightly injured by broken glass. The truck driver had left his cabin and was not injured. The train was damaged severely by the rear shaft of the trailer over a length of approximately 40 m. The truck and the trailer were also severely damaged. They were found again at the same side of the railway track, that is, the side of the warehouse.

Event Number 2 Event Numbers 1 and 2 happened at the same railroad crossing (see Figure 5.2) [6]. A truck and trailer rode on the Ochtruper Strasse in the direction of the town center on August 5, 2013. Shortly before 11.20 h, the driver stopped the combination when the truck had not yet completely passed the railroad. The driver had wanted to move backward into Presterkamp to deliver the load to the warehouse

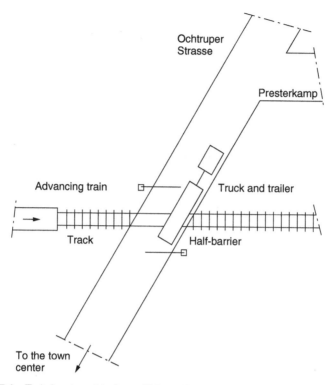

Figure 5.1 Train/truck and trailer collision – Event Number 1.

at the end of Presterkamp. The half-barriers then closed automatically because a train approached. One half-barrier came down between the truck and the trailer. Next, the train hit the combination and separated the truck from the trailer. They were found again at different sides of the railway track. The train driver and 14 passengers were slightly injured by broken glass. The truck driver had left his cabin and was also slightly injured. Both the train and the combination were damaged severely.

Additional Remarks The drivers of the two truck/trailer combinations stopped on a railroad crossing in order to be able to maneuver their combinations. That is not allowed. A railroad crossing should be passed at a moderate speed if the light signals and acoustic signals are not active.

The railroad crossing and the warehouse are very close to each other. The situation is prone to human mistakes.

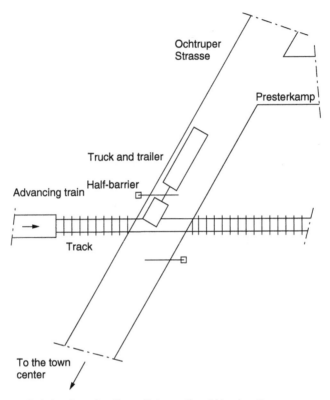

Figure 5.2 Train/truck and trailer collision – Event Number 2.

5.2.4 Derailment at Wetteren in Belgium in 2013

Event A freight-train rode on Line Number 53 from Mechelen to Schellebelle in Belgium on May 4, 2013. Schellebelle is the name of both a small community near Wetteren and a junction of the Belgian railway system. There is a switch at Schellebelle junction enabling a train riding on Line 53 from Mechelen to Schellebelle to continue its journey on Line 50 from Brussels to Port of Ghent. Port of Ghent is Gent's sea harbor. It was the intention that the train concerned would continue its trip on Line 50 in the direction of Port of Ghent. The train consisted of 2 engines and 13 wagons. Several wagons contained acrylonitrile, which is a liquid at atmospheric pressure. Its boiling point at atmospheric pressure is 77.3 °C. The compound is inflammable and toxic. The train had a speed over 80 km h^{-1} when it approached the switch mentioned at Schellebelle shortly before 02.00 h. The train passed a sign instructing the train driver to reduce the speed of the train. However, the train driver did not adjust the speed.

When he next saw a sign indicating that the speed should be 40 km h^{-1}, he activated the brakes of the train. Six wagons derailed subsequently in the switch at Schellebelle junction. The contents of three of these damaged wagons took fire. They contained acrylonitrile. The fire-brigade used water to extinguish the fire. The water used entrained liquid acrylonitrile into the local sewer system. An aspect is that acrylonitrile is soluble in water; a saturated aqueous solution of the compound contains 7.3% by weight of the material at 20 °C. The sewer system of Wetteren became polluted with acrylonitrile. Vapors were resorbed and caused the death of one man aged 64, whereas 49 people experienced serious health problems. Totally, 600 people were evacuated.

Concluding Remarks   The train driver failed to adapt the train's speed. It is improbable that this accident would have taken place in, for example, The Netherlands, where a train is automatically brought to a halt if the train driver fails to adjust the speed; see Section 3.2.3.

5.2.5   Derailment at Santiago di Compostela in Spain in 2013

Event   A fast train carrying passengers was on its way from Madrid to Ferrol in Spain on July 24, 2013. The train consisted of a front power car, eight cars, and a rear power car. Shortly before Santiago di Compostela and at approximately 20.45 h, the train had a speed of 184 km h^{-1}. Signals were given to the driver to reduce the speed to 80 km h^{-1}. The reason for the requested and required reduction of the speed is the fact that there was a bend in the railway ahead. However, the driver did, at that point in time, not adjust the speed. At a later point in time, he activated the brakes and reduced the speed to 153 km h^{-1}. The train had then already entered the bend and derailed. A total of 222 people were on board the train, of which 79 died. More than 120 people were injured, of which 20 seriously. The driver was also injured.

Concluding Remarks   See the concluding remark of the previous section (Section 5.2.4). It is again the train driver missing the signal to reduce the speed. A human failure! It is known that people tend to make mistakes. Designs should therefore include features to correct the mistakes.

5.2.6   Derailment at Port Richmond, Philadelphia, Pennsylvania, USA in 2015

Event   A train carrying 238 passengers and 5 crew members was on its way from Washington DC to New York City on May 12, 2015 [7]. The train consisted of one locomotive and seven cars. It departed Philadelphia's 30th Street Station at about 09.10 h p.m.. It traveled at a speed of 171 km h^{-1} when it entered a left curve in the Port Richmond neighborhood of Philadelphia. The train driver then applied the emergency brake and the train derailed and crashed at 09.23 h p.m., while it had a speed of 164 km h^{-1}. The entire train went off the track, with three cars rolling onto their sides. Eight people were killed in the accident and more than 200 people were injured, of which 11 critically.

Additional Remarks The driver of the manually controlled train was expected to reduce the speed to maximum 130 km h^{-1} on approaching the curve. The maximum prescribed speed in the curve was 80 km h^{-1}. However, these speeds were not selected. A projectile may have hit the windshield of the train's locomotive shortly before the derailment so the driver could not see the signals. This aspect is being investigated at present. A system to stop the train automatically or slow it to a safe speed, regardless of driver input, was not in place.

Concluding Remarks The train derailed when the emergency brake was activated to reduce the speed in an effort to take the bend. The driver's corrective action was too late.

See further the concluding remarks of Section 5.2.4.

5.2.7 Sinking of the Baltic Ace in the North Sea in 2012

Event The Baltic Ace was a roll-on roll-off (RoRo) carrier and had a load of 1417 automobiles [8]. It was on its way from Zeebrugge in Belgium to Finland. The ship collided with a container ship, Corvus J, in the North Sea in the evening of December 5, 2012. It was dark at the time of the collision. The Corvus J was on its way from Edinburgh in Scotland to Antwerp in Belgium. The Baltic Ace sunk quickly as a result of the collision. There were 24 men on board, of which 13 were saved and 11 men lost their lives. The Corvus J had suffered minor damage and could continue its journey to Antwerp.

Roll-on Roll-off Carriers and Ferries The Baltic Ace was 148 m long and 25 m wide. It had eight car decks over the full length of the ship and a capacity of 2000 automobiles. Whereas ships like the Baltic Ace are effective for the transport of automobiles, they suffer from a relatively low stability. The ships can capsize quickly when water enters through a hole in the hull.

An accident resembling the accident of the Baltic Ace occurred in 2002. The RoRo carrier Tricolore collided with the container ship Kariba in The Channel. The Tricolore sank quickly, whereas the Kariba could continue its journey.

In 1987, the RoRo ferry The Herald of Free Enterprise capsized quickly after having left the harbor of Zeebrugge in Belgium. The cause was that water entered the ship through doors that had been left open.

In 1994, the RoRo ferry Estonia capsized quickly in the Baltic Sea. The cause was that, in bad weather, water entered the ship when the doors gave in.

Concluding Remarks The stability of most RoRo ships is poor because the main part of these ships is above sea level.

The stability of a ship such as the Baltic Ace could be improved by the installation of bulkheads. Watertight doors in the bulkheads would allow the passage of cars. An additional possibility would be the provision of a double hull.

The present design of RoRo carriers and ferries is a vulnerable design from a safety point of view.

5.2.8 Aerotoxic Syndrome

Event Members of cabin crews of airplanes state that they suffer from health problems caused by the quality of the cabin air. Their immediate symptoms are headache, vision disorders, and signs of paralysis. Chronic problems of the nervous system are experienced. The problems concern, e.g. the legs, arms, and hands, the memory, and the coordination.

The Alleged Cause of the Health Problems Commercial jet airplanes were introduced in 1960s. The airplanes fly at a height of, e.g. 10–12 km. The ambient air pressure is about 0.2 bara at a height of 12 km. The air pressure in the cabin and the cockpit of airplanes is typically maintained at 0.75 bara, the ambient pressure at a height of 2400 m. Air is lost from the airplane because the cabin pressure is higher than the outside pressure. Thus, the air loss has to be made up. This is done as follows. The air needed for the combustion of the fuel in the jet engines is taken from the outside atmosphere and is compressed. Some of this air is passed on to the cabin. Thus, this make-up air has been in contact with parts of an engine. The motor contains engine oil to which an antiwear additive has been added. The direct contact between the make-up for the cabin air and the engine oil is prevented by the presence of seals. However, if the oil seals are defective, oil fumes can enter the cabin [9]. Furthermore, the seals are not gastight so that minute amounts of gaseous components resorbed by the engine oil are always entrained by the make-up for the cabin air. The gaseous components specifically in focus are resorbed by the antiwear additive. Motor engine oil typically contains 1–5% by weight of this material [10]. The chemical name of the additive is tricresyl phosphate (TCP). The additive does not consist of an unambiguously defined chemical compound but is a mixture of isomers of TCP. The antiwear additive is a neurotoxin. The threshold limit value (TLV)–time-weighted average (TWA) value has been set at $0.1 \, \text{mg m}^{-3}$. The TWA indication concerns the average exposure on the basis of an 8 hours per day, 40 hours per week working schedule.

Additional Remarks   The Boeing 787 Dreamliner takes the air for the cabin directly from the atmosphere. Thus, the make-up air does not come into contact with engine parts.

The Concorde also took the air for the cabin directly from the atmosphere. Again, the make-up air did not come into contact with parts of engines.

There are indications that there is a relationship between the health problems of members of the cabin crews of airplanes and the presence of TCP in the cabin air. However, the relationship has not yet been unambiguously established.

Concluding Remarks   In retrospect, it would have been better if the cabin air supply system selected for the Boeing 787 Dreamliner and the Concorde would have been chosen for all commercial jet airplanes in the 1960s. Thus, the occurrence of health problems caused by inhalation of TCP would have been made impossible. In Section 3.2.5, it is recommended that engineering students in the first year acquire knowledge of engineering in general. It is also stipulated that, in general, it has advantages for engineers to also have knowledge of other fields than their own field. The section is on a burning battery in a Boeing 787 Dreamliner in 2013. The aeronautical engineers making the design for the make-up of the cabin air of jet airplanes could have had a feel for the toxicity of the antiwear additive and the possibility of cabin air contamination.

This case again illustrates that safety, including toxicity items, has to be taken into account in the design stage.

5.3 SOCIETY

5.3.1 Death in a Container for Used Clothes at Hannover in Germany in 2012

Event Kristian Serban died at Hannover in Germany on February 9, 2012 [11]. He had the Rumanian nationality, was 23 years old, and worked in Germany as a building worker. He tried to take clothes out of a container for used clothes and got stuck in the opening of the container. His death was caused by asphyxiation. At the time of his death, the ambient temperature was −13 °C.

Additional Facts The container is built to prevent thefts. It is possible to push a lid and then introduce the used clothes through a 20-cm slot. Usually, the used clothes are brought to the container in plastic bags. However, Kristian Serban pushed the lid and tried to collect used clothes from the container. He then got stuck with his head in the container.

Kristian Serban and his uncle, also of Rumanian nationality, rescued another Rumanian from the same container 1 week earlier.

Concluding Remarks Everybody has seen people grabbing in garbage cans to collect items. It should not be possible that those people can manage to get into a container to find themselves trapped in the container. Containers should be designed to allow the passage of clothes into the container. At the same time, they should, normally speaking, not allow the passage of humans. I am sure that it is possible to design such containers and the collection of used clothes can be organized properly.

5.3.2 Death in a Restaurant at Zutphen in The Netherlands in 2014

Event A woman fell from a stairs in a restaurant at Zutphen in The Netherlands on April 14, 2014 [12]. She was heavily injured and was taken to a hospital, where she died.

Additional Facts The restaurant's wardrobe was close to a stairs leading from the ground floor to the cellar. The woman followed a waiter with the intention to indicate her coat in the wardrobe. On arriving at the wardrobe, she fell down the stairs, probably she focused on her coat and did not pay attention to the stairs.

Concluding Remarks   The accident would probably not have occurred when the locations of the wardrobe and the stairs would have been separated.

5.3.3   Traffic Accident at Raard in The Netherlands in 2013

Event   A car hit a group of people at Raard near Dokkum in The Netherlands shortly after 01.00h on January 1, 2013 [13, 14]. One man died in a hospital on the same day and 16 people were injured. Four people were seriously injured.

Additional Facts   A group of approximately 40 people from Raard had come together to celebrate the New Year. A fire had been lighted on the occasion by the side of a public road. Part of the group stood on the public road. Street lighting was not present. It rained at the time of the accident. The driver did not hit the people on purpose. She did not drive too fast. She had neither consumed alcoholic drinks nor was she under the influence of medicines or drugs. Her car appeared to be in working order. It is surmised that the driver did not notice the group because her sight was restricted because of the fire and the rain. The group of people celebrating the New Year did not notice the approaching car either because they paid attention to the fire, were talking with each other, and the wind was blowing.

Coming together to celebrate the New Year at this location was a tradition at Raard.

Concluding Remarks   A larger distance between a fire and a public road would have been safer. That would call for a different location of the fire. Closing off the public road at the same location would have been an alternative.

5.3.4   Accident at a Soccer Match at Eindhoven in The Netherlands in 2013

Event   A match between two teams of professional soccer players took place at Eindhoven in The Netherlands on January 18, 2013. One of the players was sent off by the referee because of an incorrect tackle. The player concerned, being on his way to the dressing room, punched a wire glass window in anger and became seriously injured at his arm.

Additional Facts   The soccer player concerned had not been able to play for 8 months due to injuries. He was disappointed by the referee's decision in the match in which he was able to rejoin the team after his relatively long absence.

Concluding Remarks It is known that soccer matches arouse emotions. A player, who was sent off the soccer field, should, on his way to the dressing room, not meet with wire glass windows or wire glass doors. It is known that wire glass windows can cause more harm to humans than wooden doors, iron doors, and stone walls.

5.3.5 A Gust of Wind at Delden in The Netherlands in 2013

Event A youth team of a soccer club at Delden in The Netherlands came together at Delden at the end of the season on June 22, 2013 [15]. An inflatable castle was one of the attractions at this party (see Figure 5.3). A gust of wind lifted up the moon bounce and threw it upside down. Several children fell out of it. Three children were injured.

Additional Facts The inflatable castle had not been anchored. And 2013 was the third year the club used the moon bounce. Neither incidents nor accidents had happened in the previous years. The inflatable castle was the property of a local school. The club had borrowed it from the school.

Figure 5.3 An inflatable castle. *Source:* Courtesy of JB-Inflatables B.V., Meppel, The Netherlands.

Concluding Remark It is known that inflatable castles and similar items should be anchored.

5.3.6 Boy Falls into Water Basin at Hengelo (O) in The Netherlands in 2013

Event A boy of 8 years old fell into the left part of an M-shaped water basin at Hengelo (O) in The Netherlands in July 2013 (see Figure 5.4) [16]. The water basin drained off into a drain via a pipe having a diameter of 50 cm and a length of 12 m. The height difference between the basin and the drain was about 1 m. The boy was entrained by the water flow through the pipe and emerged in the drain unharmed.

Additional Facts The water basin did not have a fence. A fence consisting of horizontal stainless cords had been present; however, the cords had been stolen. There was no grating between the basin and the drain.

The basin and the drain are part of the local water economy. They are also part of a recreational area.

Figure 5.4 Drain of a water basin. *Source:* Courtesy of de Persgroep Nederland BV, Amsterdam, The Netherlands.

The boy of 8 years old was accompanied by a boy of similar age. There was no senior supervision. The boys were attracted by the swirl in the basin.

Concluding Remark   A fence and a grating for the drain would have been appropriate.

5.3.7   Damaged Cow Teats at Losser in The Netherlands in 2009

Event   A milking machine was used at a farm at Losser in The Netherlands [17]. Normally, an agent is administered to the teats of the cows to take care of the teats after milking. However, instead of that agent, a cleaning agent for the milking machine was brought into contact with the teats in the evening of December 18, 2009. The cleaning agent was an aqueous solution of sodium hydroxide. The mistake was noticed the next morning. By then, the cleaning agent had caused serious chemical burns of the teats and of the udders and 61 out of a total number of 67 cows had to be slaughtered.

Additional Facts   Both the agent for the teats and the cleaning agent had been delivered to the farmer in the same blue jerrycans. The jerrycans containing the agent for the teats and the jerrycans containing the cleaning agent were labeled differently. The jerrycans containing the cleaning agent could be connected to the milking machine in the same manner as the jerrycans containing the agent for the teats. The farmer had exchanged the jerrycans containing the agent for the teats for the jerrycans containing the cleaning agent by mistake. Aqueous solutions of sodium hydroxide cause chemical burns.

Concluding Remarks   Different colors for the jerrycans containing the two agents would have been better. This practice is, for instance, well known in the chemical industry. Furthermore, it should not be possible to connect the jerrycans containing the cleaning agent directly to the line leading to the intermediate storage of the agent for the teats. This practice is also observed in the chemical industry. As a matter of fact, this has been changed in the stock farming.

REFERENCES

[1] de Volkskrant, Amsterdam, The Netherlands, August 1, 2014, pp. 1, 3 (in Dutch).
[2] NU.nl (Internet), September 19, 2009 (in Dutch). Type in Google: Rampbus Spanje reed veel te hard.

[3] Kraske, M. and Röbel, S. (2003). Blackout at Gold Coast. *Der Spiegel* 57 (20): 122–123 (in German).

[4] Spiegel Online, May 20, 2014 (in German). Type in Google: Buskatastrophe am Plattensee.

[5] Federal Railway Office (2011). Final report. Dated May 5, Bonn, Germany (in German).

[6] Federal Railway Office (2013). Final Report 60 – 60uu2013–08/00043. Dated September 13, Bonn, Germany (in German).

[7] Wikipedia (2015). *2015 Philadelphia Train Derailment.*

[8] NRC Handelsblad, Rotterdam, The Netherlands, December 6, 2012, p. 3 (in Dutch).

[9] Winder, C. and Michaelis, S. (2005). *Heidelberg Environmental Chemistry*, Vol. 4, Part H, pp. 211–228. Berlin and Heidelberg, Germany: Springer-Verlag.

[10] Furlong, C.E. (2011). Exposure to triaryl phosphates: metabolism and biomarkers of exposure. *Journal of Biological Physics and Chemistry* 11: 165–171.

[11] Schrep, B. (2012). Trapped. *Der Spiegel* (June/July), 66 (26): 26–34.

[12] De Twentsche Courant Tubantia, Enschede, The Netherlands, April 18, 2012, p. 4 (in Dutch).

[13] NRC Handelsblad, Rotterdam, The Netherlands, January 2, 2013, p. 3 (in Dutch).

[14] Director of Public Prosecutions (2013). No prosecution traffic accident at Raard. *Press Information* (16 April), Leeuwarden, The Netherlands (in Dutch).

[15] De Twentsche Courant Tubantia, Enschede, The Netherlands, June 25, 2013, pp. 2–3 (in Dutch).

[16] De Twentsche Courant Tubantia, Enschede, The Netherlands, July 18, 2013, pp. 4–5 (in Dutch).

[17] De Twentsche Courant Tubantia, Enschede, The Netherlands, January 29, 2010, pp. 27–28 (in Dutch).

6

DESIGN WITH AMPLE MARGINS

6.1 INTRODUCTION

Making designs with ample margins is recommended in this chapter. The need to pay attention to safety aspects is discussed in Chapter 4. In Chapter 5, it is advised to make accidents and incidents virtually impossible. It was sometimes an arbitrary choice in which of these three chapters a specific case would fit in best.

First, transport receives attention in Section 6.2. Road and rail transport are followed by a bob-run accident in Canada in 2010. Space travel and air travel come next. The accidents with the space shuttles Challenger in 1986 and Columbia in 2003 are described in this section.

A mine accident at Lengede in Germany in 1963, the collapse of a terminal at Roissy Airport at Paris in 2004, and the escape of a gorilla from a zoo at Rotterdam in The Netherlands are described in Section 6.3, which concerns society.

Safety in Design, First Edition. C.M. van 't Land.
© 2018 John Wiley & Sons, Inc. Published 2018 by John Wiley & Sons, Inc.

6.2 TRANSPORT

6.2.1 Coach Accident in the Sierre Tunnel in Switzerland in 2012

Event   At 21.15 h on March 13, 2012, a Belgian coach crashed in the Sierre tunnel in Switzerland. The coach was on its way to bring school children and their attendants back to Belgium. The school children had had a holiday of 9 days, in which they also had practiced skiing. There were 52 people on board the bus, of which 22 children, 4 attendants, and 2 drivers died in the crash. The bus had got off the road and had collided head-on with a wall of a lay-by, and that wall was perpendicular to the road direction (see Figure 6.1).

Additional Facts   The Sierre tunnel is near Sierre in the Swiss canton of Valais. The tunnel was opened in 1999 and consists of two tubes of 2460 m in length. Thus, it is a unidirectional tunnel. There is a European Directive on the minimum safety requirements for tunnels in the trans-European road network [1]. Switzerland is not a member of the European Union; however, it has committed itself to follow this Directive. The Directive contains instructions for lay-bys in tunnels, however, only for bidirectional tunnels.

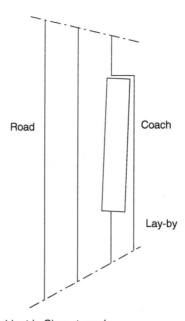

Figure 6.1   Coach accident in Sierre tunnel.

Furthermore, the Directive does not contain instructions concerning the angle between the end walls of lay-bys and the road direction.

Concluding Remarks   End walls of lay-bys of roads in tunnels should have an angle with the road direction of, e.g. 30°. Vehicles can get off the road and on the lay-by for various reasons, e.g. simply by mistake. If that is the case, they should be treated as gently as possible by deflection. The Swiss authorities have installed a crash barrier in the lay-by in question after the accident. That crash barrier has an angle of less than 30° with the road. This provides a good safety margin to mitigate the effects of collision.

6.2.2   Accident with a Bus at Almelo in The Netherlands in 2003

Event   On December 12, 2003, a bus driver was caught between the front doors of his bus at Almelo in The Netherlands when he left the bus in the bus house [2]. He was killed by the force of the closing doors.

Additional Facts   The bus driver had switched off the bus operational system and wanted to leave the bus through the front doors. The interlocking of the bus system saw to the violent closing of the front doors 4 s after the bus operational system had been switched off. It had been taken into consideration that 4 s should be the ample time for the bus driver to leave the bus after he had switched off the bus system.

Concluding Remarks   Immediately after the accident, Connexxion, the responsible bus company, ordered bus drivers to leave their buses through the exits in the middle part of the bus or at the rear of the bus. Those doors were safeguarded; that is, they closed slowly and opened automatically when experiencing a resistance. Subsequently, the company had changed the interlocking of the front doors of their buses within weeks. The front doors closed slowly after this modification and opened automatically when experiencing a resistance. The original margin of 4 s to leave the bus by the front door was too small and the violent closing of that door was unsafe.

6.2.3   Accident in a Cable Railway at Kaprun in Austria in 2000

Event   At approximately 09.00 h on November 11, 2000, a carriage of a cable railway took fire in a tunnel at Kaprun in Austria. The carriage was moving to the mountain station. A total of 150 people from the carriage died in the accident. The driver and a passenger in a train near the mountain station and three people in the mountain station died as well. Their death was caused by smoke.

Additional Facts   The cable railway transported skiers through a tunnel having a length of approximately 3.3 km. The carriage took fire after it had advanced approximately 600 m in the tunnel. Twelve passengers escaped from the fire by leaving the carriage and walking downward. Other passengers left the carriage and walked upward. They lost their lives because the flames and the smoke also moved upward.

The fire started by the ignition of a hydraulic oil. The hydraulic oil was in use for the braking system. The fire started in the driver's cabin at the rear of the train. The train had driver's cabins both at its front and its rear. The driver's cabins were integral parts of the train. At the time of the accident, the driver's cabin at the front was manned, whereas the one at the rear was not manned. Electric air heaters had been installed by the operator of the cable railway in the driver's cabins to warm the air in the cabin. A line containing hydraulic oil for the braking system was close to the air heater in the driver's cabin at the rear of the train and leaked or started to leak. The oil's trade name was Mobil Aero HFA. The hydraulic oil was inflammable.

Concluding Remarks   Placing an electric heater and a line containing an inflammable liquid close to each other should have been avoided. Reference is made to the contents of Section 4.2.3. Having an inflammable car refrigerant is not recommended. A remark made in Section 4.2.5 is also applicable in this case. Knowledge acquired by companies manufacturing cars would have been useful in this case.

6.2.4   Flashing Red Lights for Rail Transport

The safeguarding of Dutch railway transport has been described in Section 3.2.3. The safeguarding system has two shortcomings. One of these two shortcomings is that the system is not active when the train speed is lower than 40 km h^{-1}. Thus, trains having a velocity lower than 40 km h^{-1} have passed stop signs and that has caused accidents. Schijve [3] suggests to replace the stop signs emitting red light continuously by flashing red lights. Railroad crossings are equipped with flashing red lights. Car drivers making an emergency stop can activate flashing lights of their car. If implemented, the author expects a decrease in the number of times a stop sign is overlooked by a train driver in The Netherlands per annum.

6.2.5 Luge Accident at Whistler in Canada in 2010

Event   On February 12, 2010, Georgian luger Nodar David Kumaritashvili suffered a fatal crash during a training run for the 2010 Winter Olympics competition at Whistler in Canada [4].

Additional Facts   The luger was fatally injured at the Whistler Sliding Centre when he lost control in the final turn of the course and was thrown off his luge and over the sidewall of the track, hitting an unprotected steel support pole. His speed was nearly 90 miles per hour. The speeds of the lugers on the course are maximum before the final turn.

Concluding Remarks   Lugers can get off the track for various reasons. If that is the case, he or she should be treated as gently as possible by means of deflection. It is standard practice in ski resorts to shield steel support poles. After the accident, a high wooden wall was built at the location of the accident. Other measures to prevent further accidents at the 2010 Winter Olympics were taken as well.

6.2.6 Concorde Accident at Paris in 2000

Event   In the afternoon of July 25, 2000, a Concorde airplane crashed at Paris. It fell on a hotel at Gonesse after having taken off from Roissy-Charles de Gaulle Airport. A severe fire broke out under the left wing when the aircraft was still on the runway. The fire caused a loss of thrust of both engines attached to the left wing (Numbers 1 and 2). All 109 people on board the airplane and 4 people in the hotel lost their lives. The airplane had been in the air for approximately three quarters of a minute. Both the aircraft and the hotel were destroyed.

Probable Cause of the Accident   The front right tire of the left landing gear having four tires (Tire 2) ran over a strip of metal, which had fallen from a different airplane, and was damaged [5]. Debris was thrown against the left wing and it caused the rupture of a fuel tank in the wing (Tank 5). The flow rate of leaking fuel has been estimated as several dozens of kg s^{-1}. The leaking fuel ignited by an electric arc in the landing gear or through contact with hot parts of an engine. The speed of the aircraft at that point in time was 280 km h^{-1}. That means that the aircraft was committed to take-off. The fire caused a loss of thrust on Engine 2. The crew shut down Engine 2 following an engine fire alarm. The landing gear would not retract. Next, Engine 1 lost thrust, and, subsequently, the aircraft's angle of attack and bank increased sharply. Finally, Engines 3 and 4 on the right wing lost thrust by a combination of deliberate selection of idle by the crew and by a surge due to excessive air flow distortion. This allowed aircraft bank to be reduced. The Concorde then crashed practically flat, destroying the hotel. It was then immediately consumed by a violent fire.

Concorde Features The Concorde was an airplane designed to fly with a speed in the range of 2000–2200 km h^{-1} at an altitude of approximately 17 km. The aircraft is depicted in Figure 6.2. This forced designers to attach two engines to each wing. The two engines attached to a wing were installed side by side. This design implies that a problem with an engine may affect the engine next to it. Furthermore, the engines are close to fuel tanks in the wings. Compare this design to that of, e.g. the Airbus 340 or the Boeing 747. Both the distance between the engines and the distance between the engines and the fuel tanks in the wings are relatively large for those airplanes.

The wings of a Concorde were more vulnerable to the impacts of foreign objects than the wings of, e.g. a Boeing 747 or an Airbus A340. For

Figure 6.2 Concorde aircraft. *Source:* Courtesy of Mr. Henk Heiden, Oosterhout, The Netherlands.

example, the skin thickness of Tanks 2, 5, and 6 was only 1.2 mm [6]. These tanks were damaged in an incident at Washington in 1979.

The speed with which a Concorde took off and landed was approximately 400 km h^{-1} [7]. Compare this to the speed with which a Boeing 747 takes off and lands, that is, approximately 325 km h^{-1}. The high speed with which a Concorde took off and landed implied a heavy load for the tires.

Concorde History The Concorde resulted from a collaboration between France and Great Britain. A total of 20 Concordes have been built. British Airways and Air France together operated 14 airplanes. The commercial exploitation started in 1976. The number of flying hours before the accident at Paris in 2000 was 300 000. Both the English and the French Concordes were grounded within a month after July 25, 2000, the date of the accident. Approximately 1 year later, after the implementation of improvements, both operators were allowed to fly again with the aircraft. However, Air France stopped flying with the Concorde in March 2003 and British Airways in October 2003. Both companies mentioned the decrease in the number of passengers as the reason.

Previous Events Concerning Tires BEA (Bureau d'Enquêtes et d'Analyses pour la Sécurité de l'Aviation Civile) mentions 57 cases of tire bursts/deflations for the Concordes prior to the date of the accident at Paris on July 25, 2000 [8]. The number 57 has been assessed as the number of events for which information from at least two different sources had been found or for which reports or detailed information exist. Twenty-one further events were notified by a single source, but no reports or detailed information exist for them. Mention of damage to the structure of the tanks was not made in any of these 21 events. Out of the 57 events, Air France experienced 30 and British Airways 27. Twelve out of these 57 events had structural consequences on the wings and/or the tanks, of which six led to penetration of the tanks. The latter six incidents occurred between 1979 and 1994. Only one out of these six incidents at which tanks were penetrated resulted in a fuel leak. This occurred at the aforementioned incident at Washington in 1979. The fuel leak from all of the penetrations at Washington in 1979 was 4 kg s^{-1}.

None of the 57 events identified showed rupture of a tank, a fire, or a significant simultaneous loss of power on two engines.

Only one case of tank penetration by a piece of tire was noted. Metal parts originating from, e.g. wheel rims punctured the tanks in the remaining five cases in which tanks were penetrated.

Several measures were taken in the course of the years, e.g. both Air France and British Airways stopped using retread tires. However, there were still three tire bursts in 2000 before July 25, 2000.

Concluding Remarks   The deflating/bursting of tires of the Concorde during take-off and landing was not considered a safety risk between 1976 and July 25, 2000. That point of view was held when serious consequences did not result from the incidents. After the accident at Paris on July 25, 2000, it was considered a safety risk. An important measure to prevent tire bursts/deflations was the introduction of the NZG-tire (near zero growth) made by Michelin. A further important measure was the reinforcement of several fuel tanks by means of a Kevlar lining. Kevlar is a material used for, e.g. the manufacture of bulletproof vests. Further measures were implemented after July 25, 2000.

6.2.7   Space Shuttle Challenger Accident in 1986

Event   The Space Shuttle Challenger accident occurred on January 28, 1986, when it broke apart 73 s into its flight. Its Mission Number was STS-51-L (STS stands for Space Transport System). The spacecraft disintegrated over the Atlantic Ocean, off the coast of Central Florida. The disintegration led to the death of all seven crew members.

Space Shuttle   Space Shuttles were launched from Kennedy Space Center at Cape Canaveral in Florida as from 1981 [9]. The program was stopped in 2011. A Space Shuttle consisted of four parts (see Figure 6.3). First was an Orbiter looking like an airplane having two short wings. Second and third were two Boosters containing solid fuel. They provided the major part of the thrust during the first part of the launch. They were empty after approximately 2 min, were pushed off, and were parachuted down into the Atlantic Ocean. The Boosters were reused. Fourth was an External Tank (ET) containing liquid oxygen at −183 °C and liquid hydrogen at −235 °C. The three main engines of the Space Shuttle were activated 7 s before the launch. They received both liquid hydrogen (fuel) and liquid oxygen from the ET. The ET was empty after approximately 8 min and was subsequently jettisoned from the Orbiter. It then burnt up in the earth's atmosphere. The main engines stopped functioning when the ET was empty and were not used anymore during the particular flight. Launching still lasted approximately 2 min after the ET had been jettisoned from the Orbiter. Engines smaller than the main engines then provided the further required thrust.

The ET was insulated with foam to prevent ice formation on the ET.

Figure 6.3   STS-1 (Columbia) at liftoff. *Source:* Courtesy of NASA, USA.

O-rings   Each of the two Boosters consisted of a number of separate segments joined together and sealed by O-rings. The segments were shipped from their manufacturer and assembled at the Kennedy Space Center. The Challenger accident occurred when hot gases burned through an O-ring and seal in the aft joint on the left Booster. The Presidential Commission on the Space Shuttle Challenger Accident published a report. The title of Chapter 6 is "An Accident Rooted in History". In the Findings of this chapter, the following text can be read:

> The Commission has concluded that neither Thiokol nor NASA responded adequately to internal warnings about the faulty seal design. Furthermore, Thiokol and NASA did not make a timely attempt to develop and verify a new seal design after the initial design was shown to be deficient. Neither organization developed a solution to the unexpected occurrences of O-ring erosion and blow-by even though this problem was experienced frequently during the Shuttle flight history. Instead, Thiokol and NASA management came to accept erosion and blow-by as unavoidable and an accepted flight risk.

The title of Chapter 4 of the aforementioned report is "The Cause of the Accident". The Conclusion of this chapter reads as follows:

In view of the findings, the Commission concluded that the cause of the Challenger accident was the failure of the pressure seal in the aft field joint of the right solid rocket motor. The failure was due to a faulty design unacceptably sensitive to a number of factors. These factors were the effects of temperature, physical dimensions, the character of materials, the effects of reusability, processing, and the reaction of the joint to dynamic loading.

Concluding Remarks The margins of the Space Shuttle with respect to O-ring erosion and blow-by were too small. The joints were redesigned after the accident.

6.2.8 Space Shuttle Columbia Accident in 2003

Event The Space Shuttle Columbia tried to return from a mission on February 1, 2003. Its Mission Number was STS-107. It disintegrated over Texas during re-entry into the earth's atmosphere. This resulted in the death of all seven crew members.

Space Shuttle See Section 6.2.7 and Figure 6.3.

High Temperatures on Re-entering the Earth's Atmosphere The orbital speed of Columbia was approximately $28\,000\,\mathrm{km\,h^{-1}}$. The Commander and the Pilot of Columbia used the two Orbital Maneuvering System engines to slow down Columbia, to leave the orbit, and to reenter the earth's atmosphere. This occurred over the Pacific Ocean at a height of $122\,\mathrm{km}$. Air consists of nitrogen and oxygen mainly. Columbia collided with these two types of molecules and that produced friction heat. During such a descent, wing leading-edge temperatures rise to values probably exceeding $2760\,^{\circ}\mathrm{C}$ [10]. The Space Shuttle was protected against high temperatures by a thermal protection system (TPS). To this end, reinforced carbon–carbon (RCC) panels were installed on the leading edges of the wings. However, shortly after the launch, a suitcase-size piece of thermal insulation foam had broken off from the ET and struck the leading edge of Columbia's left wing, damaging an RCC panel. The piece of foam came off an area where the Orbiter attaches to the ET. Part of the hit RCC panel left the Columbia when it was in orbit. Thus, Columbia was, at this position, no longer protected against the high temperatures caused by the compression of the gas on re-entry of the earth's atmosphere and that caused the destruction of the internal wing structure and ultimately of the vehicle.

Foam Shedding   The shedding of ET-foam had a long history [11]. Damage caused by debris has occurred on every Space Shuttle flight, and most missions have had insulating foam shed during ascent. One debris strike in particular foreshadows the STS-107 event. On December 2, 1988, Space Shuttle Atlantis was launched on STS-27R. During the ascent, shed insulating foam had knocked off a tile of the TPS, exposing the Orbiter's skin to the heat of re-entry. The structural damage was confined to the exposed cavity left by the missing tile, which happened to be at the location of a thick aluminum plate covering an L-band navigation antenna. Probably, the thick aluminum plate prevented the occurrence of further serious damage. Atlantis suffered further damage during the ascent on December 2, 1988.

Changes to Space Shuttle operations were implemented after the Columbia accident in 2003. On-orbit inspections of the TPS were organized to detect damage. In case of irreparable damage, a rescue mission could be sent. A further decision was that, in principle, missions would be flown to the International Space Station (ISS) only.

Space Shuttle Discovery was launched on July 26, 2005, on the "Return to Flight" mission STS-114. The mission was successful. However, a piece of foam was shed from the ET. The debris did not hit the Orbiter. The size of this piece of foam was comparable to the size of the piece of foam that hit the Columbia. The foam did not come from the same location as that in the case of the Columbia [12]. Due to this foam shedding, the next shuttle flight did not take place until July 2006. The second "Return to Flight" mission, STS-121, was launched on July 4, 2006, and was successful. However, more foam was shed than expected [12].

The foam shedding at the two "Return to Flight" missions demonstrated that foam shedding remained a problem. It had not been possible to make the Orbiter more resistant to debris strikes either.

Concluding Remarks   The margins of the launching of Space Shuttles with respect to foam shedding were too small.

The thought may come up why foam was used in the first place. Providing the ET with a double hull and creating vacuum in the concentric space would also have provided insulation and would not have caused shedding problems. However, such a design would have added mass to the ET, and at complicated spots foam would still be needed.

6.2.9   Air France Flight AF 447 Accident in 2009

Event   An airplane type Airbus A330 operated by Air France took off from Rio de Janeiro in Brazil on May 31, 2009. The aircraft was bound for

Paris in France. It experienced problems over the Atlantic Ocean in the early morning of June 1. The problems could not be controlled by the pilots and the airplane hit the surface of the ocean. The impact resulted in the death of all 216 passengers and the crew of 12 people. The airplane was destroyed.

Cause of the Accident   The speed indications became incorrect at 2 h 10 min 5 s, likely due to the obstruction of the Pitot probes by ice crystals [13]. The Pitot probes are at the outside of the airplane in contact with the atmosphere. Some automatic systems were subsequently disconnected. The flight crew could not control the flight path. It plunged into the sea at 2 h 14 min 28 s.

History of the Obstruction of the Pitot Probes   Airbus was informed by 10 operators of A330 and A340 aircraft of 16 relevant incidents that occurred in cruise between February 2005 and March 2009 [14]. The manufacturer associated these 16 incidents with the failure condition manifested by a sudden reduction in several indicated speeds. Based on the data available, these incidents could be attributed to a possible obstruction of at least two Pitot probes by water or ice. Nine of them occurred in 2008 and three at the start of 2009.

An Airworthiness Review Meeting (ARM) took place in December 2008. The "Pitot icing" theme was on the agenda. Airbus presented 17 cases of temporary Pitot probe blocking that had occurred on the long-range fleet between 2003 and 2008, including nine in 2008. Airbus could not explain the sudden increase in the number of incidents.

A further ARM meeting was held on March 11 and 12, 2009. The situation concerning the Pitot probes was reviewed. It was recommended to replace the Pitot probes by improved ones. However, the replacement was not made mandatory.

The European Aviation Safety Agency (EASA) wrote a letter dated March 30, 2009, to the Direction Générale de l'Aviation Civile (DGAC). DGAC is the French Civil Aviation Authority (CAA). In this letter, the EASA organization concluded that at that stage the situation did not mean that a change of Pitot probes on the A330/A340 fleet had to be made mandatory. EASA is a European Union Agency.

Changes Made by Air France Following the Accident   The replacement of the Pitot probes by the aforementioned improved type was accelerated [15]. The replacement was completed on June 11, 2009. Subsequently, following an Airworthiness Directive issued by EASA, the

Pitot probes in Positions 1 and 3 were replaced in early August 2009 by Pitot probes made by a different manufacturer. Final remark: prior to the AF 447 accident, Air France took the initiative to have the Pitot tubes in Position 2 replaced in January and February 2009 by Pitot probes made by the different manufacturer. Unfortunately, the replacement of one probe is not enough. Both the Airbus A330 and the Airbus A340 aircraft have Pitot probes in three positions.

Concluding Remarks   The design of the Pitot probes was not a design with ample margins for the application in airplanes. Moreover, in case of failure of the probes, the flight crew had relatively little time, in the AF 447 accident less than 4 min, to make the required adjustments.

6.2.10 Turkish Airways Flight TK1951 Accident Near Amsterdam in 2009

Event   An airplane type Boeing 737-800 operated by Turkish Airlines took off from Istanbul in Turkey on February 25, 2009 [16]. The aircraft was bound for Amsterdam in The Netherlands. It experienced problems shortly before it could have reached its destination. The problems could not be controlled by the pilots. When the airplane approached a runway, it lost height and fell, 1.5 km before the runway, on a field (see Figure 6.4). A total

Figure 6.4 Crashed Turkish Airlines Flight TK1951. *Source:* Courtesy of Dutch Safety Board, The Hague, The Netherlands.

of 128 passengers were on board and 7 persons of the flight crew. The impact resulted in the death of five passengers and four of the flight crew, and 117 passengers and 3 persons of the flight crew were wounded.

Cause of the Accident There were two instruments to measure the aircraft's altitude [17]. The output of the left instrument controlled the fuel supply to the engines (autothrottle). The left instrument gave a value of −8 ft when the aircraft approached Amsterdam Schiphol Airport. This value was wrong. Wrong values had been measured at earlier occasions. The measurement led to a reduction of the fuel supply to the engines to a minimum value. The consequence was that the airplane lost too much speed. This fact was, in first instance, not noticed by the pilots. The pilots could have seen visual indications and warnings that the airplane lost too much speed and that the nose position was not right. A warning (by shaking of the stick) that the aircraft could stall was given when the altitude was 460 ft. The flight crew noticed the latter warning but did not react adequately and the airplane hit the ground shortly after this warning.

A second aspect is that the pilots of Turkish Airlines got landing instructions from the Air Traffic Control of Amsterdam Schiphol Airport that were not completely in line with the internal directions of the Air Traffic Control of Amsterdam Schiphol Airport. This fact caused a masking of the reduction of the fuel supply to the engines to a minimum value by the wrong value of the left instrument for measuring the altitude.

Finally, the approach of the landing strip had not been stabilized by the pilots. A stabilization of the approach comprises that the obligatory activities of the pilots prior to the landing have been completed when the airplane reaches a certain altitude. One of these obligatory activities is the preparation of a checklist for the landing of the aircraft. The pilots were still busy with the obligatory activities when the stick shaker signal came.

Functioning of the Instruments for Measuring the Altitude The instruments are located in the bottom of the aircraft [18]. Parts of the instrument are in contact with the atmosphere. An antenna sends radio waves to the ground. The radio waves are reflected by the ground and received by a second antenna. The instrument calculates the altitude from the time elapsed between emission and receipt of the radio waves.

History of the Problems with the Instruments for Measuring the Altitude Experienced by Turkish Airlines Turkish Airlines experienced problems with the instruments for measuring the altitude before the accident occurred at Amsterdam on February 25, 2009 [19]. The problems experienced most frequently were fluctuating and negative altitudes, the activation of the warning system of the landing gear, the cut out of the autopilots, and warnings from the ground proximity system. A total of 235 faults of the instruments for measuring the altitude of 52 Boeing 737-800 airplanes of Turkish Airlines in the period from January 2008 to February 2009 were reported in the maintenance files of Turkish Airlines. Sixteen of these 235 faults came from the airplane involved in the accident at Amsterdam on February 25, 2009. Actions to remedy the faults were taken.

Appreciation of the Problems Experienced with the Instruments for Measuring the Altitude   Turkish Airlines considered the problems as a technical problem. The operator did not see the problems as a safety risk [20]. Boeing also concluded that the problems were not a safety problem. As far as the Dutch Safety Board could ascertain, the latter conclusion was based on the reasoning that the probability of an incident under 500 ft is very small. Boeing considered the system to be more accurate at low altitudes than at high altitudes, making the probability of the occurrence of problems at low altitudes smaller than at high altitudes. Moreover, Boeing had the opinion that the pilots get a sufficient number of warnings and instructions to take timely action, to restore the situation, and to land safely [21].

Concluding Remarks   The Number 1 Recommendation of the Dutch Safety Board in its report is:

> Boeing has to increase the reliability of the system to measure the altitude by means of radio waves [22].

Turkish Airlines and Boeing hold the point of view that the problems with the instruments for measuring the altitude are a safety risk since the accident at Amsterdam on February 25, 2009.

Accidents with Flights AF 447 and TK1951 indicate that people are safetywise not 100% reliable to take proper action in case of emergencies.

6.3 SOCIETY

6.3.1 Mine Accident at Lengede in Germany in 1963

Event   The accident occurred in the iron ore mine Lengede-Broistedt on October 24, 1963. Approximately $700\,000\,m^3$ of water ran from a pond, which was part of the facilities and adjacent to the mine, into the mine and filled it from 100 m below ground level up to 60 m below ground level. The maximum depth of the mine was 100 m. A total of 129 men were trapped in the mine, of which 79 managed to get out in the first couple of hours. Seven men were rescued on October 25. By means of a drilling action, three men were rescued on November 1. A final spectacular drilling action saved the lives of 11 miners on November 7. The total number of miners who lost their lives was 29.

Additional Facts   The pond was used for clarification purposes. That means, the pond was used to accomplish a solid/liquid separation. Water that had been used in the facility, and still contained small solid particles, was transferred to the pond. The small solid particles settled in the pond. The supernatant liquid, i.e. clear water, could then be disposed of. The pond had been installed in a pit in which open-cast mining had been practiced. There were connections between the pit and the mine before it was changed into a pond. These connections had been filled before the pit had been turned into a pond.

Concluding Remarks   A separation between the pond and the mine collapsed on October 24, 1963. It has not been possible to assess whether the accident started with a collapse of a former connection between the pond and the mine or of the separation between the pond and the mine. If the distance between the pond and the mine would have been larger and if there would never have been connections between the pit, which became a pond, and the mine, a collapse would have been impossible.

6.3.2 Collapse of Terminal 2E of Roissy Airport at Paris in 2004

Event   Terminal 2E of Roissy Airport at Paris had been taken into service in June 2003 [23]. It had a length of 650 m, and the upper part, the departure hall, had an oval Cross-section. The lower part was the arrival hall. It had facilities for nine aircrafts (see Figures 6.5 and 6.6). Terminal 2E had a daring design with wide open spaces. Arches stood from one side of the departure hall to the other side of that hall, and the distance between the supports of an arch was 26.20 m. That is, the width of the departure hall was 26.20 m, whereas the length was 650 m. The width of the arches was 4 m. Reinforced concrete, steel, and glass were used as materials of construction.

A piece of concrete fell from an arch at 05.30 h on Sunday, May 23, 2004. An evacuation was organized subsequently. A part of the terminal,

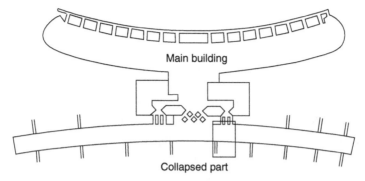

Figure 6.5 Terminal 2E at Roissy-Charles de Gaulle Airport – Top view.

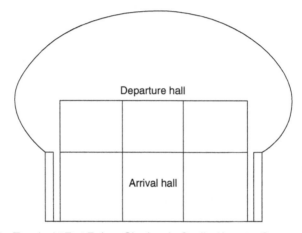

Figure 6.6 Terminal 2E at Roissy-Charles de Gaulle Airport – Cross-section.

containing the aforementioned arch, and having a length of 30 m, collapsed at 06.57 h on May 23, 2004. The ambient temperature at the time of the collapse was 4.10 °C. There were six arches in the part that collapsed. There were passenger passages through three of these arches, i.e. one passage through each arch. The accident took the lives of four people, and three people were injured.

Information   Standing in front of the terminal, in the area where the aircrafts park, the zone that collapsed was at the right-hand side of the center of the terminal. A zone symmetrical with the collapsed zone was at the left-hand side of the center. The latter zone had the same construction as the collapsed zone. Both zones have been investigated by the French

Berthier Commission (in French: La Commission Berthier). According to the Berthier Commission, the causes of the accident are as follows:

- Insufficient concrete reinforcement or improperly positioned concrete reinforcement
- A lack of mechanical redundancy, which means a lack of possibilities to transfer the load to other zones in case of local defectiveness
- A weak carrying capacity of the supporting beam
- The positioning of the steel support struts within the concrete.

Additional Information   It is the Commission's opinion that the process of the collapse of a structure having a small margin as to the carrying capacity can be explained by

- a small movement of the structure as a result of its delayed deformations related to concrete creep that, although it is normal for a concrete structure, have added to the stresses in certain susceptible points;
- the effect of cyclic temperature changes that have progressively enlarged the initial crack in the structure.

The width of the margin as to the carrying capacity of the structure was thus reduced and subsequently annulled. Thus, a minor phenomenon was sufficient to set the fatal set of events in motion. Note that the accident happened almost 1 year after Terminal 2E had been taken into service. The first event was a crack in the concrete of the inner face of an arch containing passenger passages on the line connecting the steel support struts. It probably occurred between two steel support struts. This caused the fall of a large concrete slab. The disturbance that set the fatal series of events in motion may have been the very low temperature in the morning of May 23, 2004, or the weakness of a clamp of a post.

The fall of that concrete slab has, approximately one and a half hour later, caused two interacting phenomena, which occurred almost simultaneously. These two phenomena have caused, because of the lack of redundancy of the structure, its sudden collapse. The first phenomenon was the breaking of the northern part of an arch. The second phenomenon was the breaking and the fall of the southern supporting beam.

The conclusion is that the collapse of the structure, of which the original safety margins were too narrow, has been caused by several factors instead of by one factor.

The decision to tear down and rebuild the whole part of Terminal 2E, of which a section had collapsed, was taken in 2005. A more traditional steel and glass structure was chosen. It was taken into service in 2008.

Concluding Remark   The margins of the original construction of Terminal 2E at Roissy Airport at Paris were too small.

6.3.3   Escape of a Gorilla in a Zoological Garden at Rotterdam in The Netherlands in 2007

Event   On May 18, 2007, the gorilla Bokito escaped from his residence at the zoological garden Blijdorp at Rotterdam in The Netherlands [24]. He jumped across the moat separating him and the other gorillas from the public and subsequently attacked a woman and wounded her severely. A second woman fled, fell, and broke her pelvis bone.

Characteristics of the Gorilla Residence Concerned   The width of the moat was 4 m. The Gorilla EEP Husbandry Guideline 2006 of The European Association of Zoos and Aquaria prescribes a moat width of 6 m. New gorilla residences had to adhere to this Guideline as from 2006. The gorilla residence at Blijdorp was built in 1999. The Guideline stipulated that zoological gardens were not, if applicable, obliged to increase the moat width of existing gorilla residences as long as they did not modify their gorilla residence.

Measures Taken by Zoological Garden Blijdorp After the Accident   The zoological garden had a high wooden fence built at the public side of the moat. The erection of such a wooden fence is not considered a modification. It is considered a safety measure. With the construction of a wooden fence, Blijdorp complied with the requirements of the aforementioned Guideline.

Concluding Remark   Blijdorp could have decided to build the fence when the aforementioned Guideline appeared as the specialists of the European Association of Zoos and Aquaria in 2006 considered a moat width of 4 m inadequate.

REFERENCES

[1] Council of the European Union, European Parliament (2004). Directive 2004/54/EC of the European Parliament and of the Council of 29 April 2004

on minimum safety requirements for tunnels in the trans-European road network.

[2] De Twentsche Courant Tubantia, Enschede, The Netherlands, December 15, 2003, p. 13 (in Dutch).

[3] Schijve, J. (2012). Flashing red light is better visible. *De Ingenieur* 124 (9): 64. (in Dutch).

[4] Longman, J. (2010). Quick to Blame in Luge, and Showing No Shame. *New York Times* (13 February).

[5] Bureau d'Enquêtes et d'Analyses pour la Securité de l'Aviation Civile (BEA) (2002). *Accident on 25 July 2000 at La Patte d'Oie in Gonesse (95) to the Concorde Registered F-BTSC Operated by Air France*, 14, 17, 119, 176. Paris: BEA.

[6] Bureau d'Enquêtes et d'Analyses pour la Securité de l'Aviation Civile (BEA) (2002). *Accident on 25 July 2000 at La Patte d'Oie in Gonesse (95) to the Concorde Registered F-BTSC Operated by Air France*, 95. Paris: BEA.

[7] den Hertog, R. (2002). New suspenders for the Concorde. *De Ingenieur* 114 (7): 30–33. (in Dutch).

[8] Bureau d'Enquêtes et d'Analyses pour la Securité de l'Aviation Civile (BEA) (2002). *Accident on 25 July 2000 at La Patte d'Oie in Gonesse (95) to the Concorde Registered F-BTSC Operated by Air France*, 93–97. Paris: BEA.

[9] Gehman, H.W., Barry, J.L., Deal, D.W. et al. (2003). *Columbia Accident Investigation Board Report*, 9–15. Washington, DC.

[10] Gehman, H.W., Barry, J.L., Deal, D.W. et al. (2003). *Columbia Accident Investigation Board Report*, 12. Washington, DC.

[11] Gehman, H.W., Barry, J.L., Deal, D.W. et al. (2003). *Columbia Accident Investigation Board Report*, 121–131. Washington, DC.

[12] Wikipedia (2015). Space Shuttle Columbia Disaster.

[13] Bureau d'Enquêtes et d'Analyses pour la Sécurité de l'Aviation Civile (BEA) (2012). *Final Report on the Accident on 1st June 2009 to the Airbus A330-203 Registered F-GZCP, Operated by Air France, Flight AF 447 Rio de Janeiro – Paris*, 17. Paris: BEA.

[14] Bureau d'Enquêtes et d'Analyses pour la Sécurité de l'Aviation Civile (BEA) (2012). *Final Report on the Accident on 1st June 2009 to the Airbus A330-203 Registered F-GZCP, Operated by Air France, Flight AF 447 Rio de Janeiro – Paris*, 141–143. Paris: BEA.

[15] Bureau d'Enquêtes et d'Analyses pour la Sécurité de l'Aviation Civile (BEA) (2012). *Final Report on the Accident on 1st June 2009 to the Airbus A330-203 Registered F-GZCP, Operated by Air France, Flight AF 447 Rio de Janeiro – Paris*, 215. Paris: BEA.

[16] Dutch Safety Board (2010). *Crashed During Approach, Boeing 737-800, Near Amsterdam Schiphol Airport*, 5–11. The Hague, The Netherlands: Dutch Safety Board (in Dutch).

[17] Dutch Safety Board (2010). *Crashed During Approach, Boeing 737-800, Near Amsterdam Schiphol Airport*, 47–53. The Hague, The Netherlands: Dutch Safety Board (in Dutch).

[18] Dutch Safety Board (2010). *Crashed During Approach, Boeing 737-800, Near Amsterdam Schiphol Airport*, 24. The Hague, The Netherlands: Dutch Safety Board (in Dutch).

[19] Dutch Safety Board (2010). *Crashed During Approach, Boeing 737-800, Near Amsterdam Schiphol Airport*, 49. The Hague, The Netherlands: Dutch Safety Board (in Dutch).

[20] Dutch Safety Board (2010). *Crashed During Approach, Boeing 737-800, Near Amsterdam Schiphol Airport*, 50. The Hague, The Netherlands: Dutch Safety Board (in Dutch).

[21] Dutch Safety Board (2010). *Crashed During Approach, Boeing 737-800, Near Amsterdam Schiphol Airport*, 52. The Hague, The Netherlands: Dutch Safety Board (in Dutch).

[22] Dutch Safety Board (2010). *Crashed During Approach, Boeing 737-800, Near Amsterdam Schiphol Airport*, 12. The Hague, The Netherlands: Dutch Safety Board (in Dutch).

[23] Berthier Commission (2005). *Integral Text of the Summary of the Activities of the Berthier Commission*. Paris: Conseil National des Ingénieurs et des Scientifiques de France (in French).

[24] NRC Handelsblad, Rotterdam, The Netherlands, May 23, 2007, p. 3 (in Dutch).

7

THE RISKS OF
ENCLOSED SPACES

7.1   INTRODUCTION

The reduction of oxygen (O_2) concentration in air by inert gases is a risk of enclosed spaces. A further risk is intoxication by absorption of hazardous toxic materials through the lungs. Inhalation of dust and particulate matter can also be a serious respiratory problem. The focus in this chapter is on the first two risks. Toxic materials can be classified in terms of their physiological action:

- Irritants, which includes corrosive gases/vapors that attack the mucous membrane surfaces of the body. Sulfur dioxide and chlorine are examples.
- Asphyxiants, which are substances that interfere with the oxidation processes in the body. The simple asphyxiants are physiologically inert gases, which dilute or replace the oxygen required for breathing. Dilution of air by the simple asphyxiant nitrogen (N_2) was probably the cause of the accident described in Section 7.3, the section concerning industry. Consumption of oxygen in air by fire was probably the cause

Safety in Design, First Edition. C.M. van 't Land.
© 2018 John Wiley & Sons, Inc. Published 2018 by John Wiley & Sons, Inc.

of many casualties at a fire described in Section 7.4.1, part of the section on society. Chemical asphyxiants react with an essential body function involved with the transportation of oxygen (O_2) from the lungs via red blood cells to body tissues. In such cases, asphyxiation results even though the air contains an adequate concentration of O_2. Chemical asphyxiation by carbon monoxide (CO) was the cause of the accident described in Section 7.2, the section regarding transport. A further case of chemical asphyxiation by probably both hydrogen sulfide and hydrogen cyanide is described in Section 7.4.2, part of the section on society.

- Anaesthetics and narcotics, which depress the central nervous system and lead to unconsciousness.
- Systemic poisons, which injure or destroy internal organs of the body.

7.2 TRANSPORT

Lethal accident aboard the Dutch ship Lady Irina in 2013 [1].

Event

The chief engineer of the Dutch ship Lady Irina died aboard the ship after having entered the bow thruster space in the front part of the ship on July 13, 2014. A Danish coroner established that the chief engineer had died from poisoning by carbon monoxide (CO). The ship was on its way from Archangelsk in Russia to Kolding in Denmark. It carried wood pellets.

Ship, Cargo, and Trip

See Figure 7.1. The ship has an overall length of 88 m and a maximum width of 14.4 m. Its gross tonnage is 3323. Its maximum velocity is 13.5 knots (25.0 km h^{-1}). Eight men were aboard the ship at the time of the accident. The Lady Irina regularly loaded wood pellets at Archangelsk and carried them to various European harbors, one of which is Kolding.

Detailed Description of the Accident

The chief engineer met the first chief on the bridge of the ship at approximately 19.00 h on July 13, 2014. They discussed the removal of splash water aboard the ship by a pump. He left the bridge between 19.30 and 19.45 h for the engine room. At that time, it would have taken approximately

Figure 7.1 Ship Lady Irina. *Source:* Courtesy of Dutch Safety Board, The Hague, The Netherlands.

4 h to Kolding. The captain took over from the first chief on the bridge at 20.00 h. The first chief then went to the chief engineer's cabin to discuss the work of the next day. They used to have such a meeting at that time of the day. However, the chief engineer was not present. That did not surprise the first chief; he assumed that the chief engineer was busy because of the planned arrival at Kolding. The first chief returned to the chief engineer's cabin at 21.45 h. The chief engineer still was not present. The first chief then went to the engine room; however, he did not see the chief engineer. He noticed in the engine room that the pump for the removal of splash water was running. That surprised him because the removal of splash water usually did not take longer than 20 min. The first chief thereupon went to the bow thruster space in the front part of the ship where, he assumed, the chief engineer could be present. He found that the door of that space was open. The compartment has two floors. On entering

the space, it is possible to descend a stairs leading to the lower floor. The first chief found the chief engineer lying on the lower floor. He noticed that the chief engineer did not breathe anymore and his heart had stopped beating. The first chief then went at about 22.00 h to the bridge and informed the captain and the second chief. The captain then decided to sail to the nearest harbor, i.e. Frederica in Denmark at a distance of approximately four sea-miles. The first chief ordered the second chief to get a stretcher and go to the space where the chief engineer laid. The first chief then went to the day-room to alert the crew and asked them to go to the compartment where the chief engineer was lying. He put on his working-clothes and arrived in the specific space a few minutes later to supervise the activities to remove the chief engineer from the bow thruster space. The alerted crew members had started with the reanimation of the chief engineer; however, they did not use a breathing apparatus because they thought that the chief engineer had fallen from the stairs. It did not occur to them that the air quality was the problem.

At that time, all crew members, except the captain, were involved in the rescuing operation. The group consisted of six men, i.e. the first chief, the second chief, the apprentice-engineer, the cook, and two sailors. The ship arrived at about 22.45 h at Frederica. The apprentice-engineer and a sailor thereupon left the compartment where the chief engineer was lying for the engine-room to start the rotation of the bowscrew.

The second chief left the space when it became clear that it was not possible to get the strongly built chief engineer out of the compartment with the stretcher. He went to look for a neck support and an oxygen-breathing trunk. He went for the neck support because the crew thought that the chief engineer had fallen from the stairs and he went for the oxygen-breathing apparatus because the chief engineer did not breathe anymore.

When the second chief got back, he saw that the first chief had become unconscious and was lying on the floor. One crew member tried to move him. A further crew member walked around as if he was drunk. The second chief entered and put the oxygen-breathing apparatus over the mouth of the first chief and heard that it functioned. He also requested the crew members still standing on their feet to leave the compartment. He then also left the compartment because he became unwell.

The ship moored at 22.50 h. The Danish fire-brigade got the chief engineer and the first chief out of the bow thruster space. A coroner established that the chief engineer had died because of poisoning by CO. The first chief regained consciousness after oxygen had been administered to him. Three crew members, i.e. the first chief, a sailor, and a cook, were hospitalized but could return to the ship after a couple of days.

The Cause of the Accident

Toxicological Properties of CO CO is a particularly dangerous chemical because it cannot be detected by the natural senses of the body [2]. It is toxic because it competes successfully with oxygen (O_2) for the binding sites of hemoglobin (Hb), the O_2-carrying hemoprotein in the blood of mammals. The affinity of hemoglobin for CO, calculated from the pressure of CO required for half-saturation, i.e. [Hb] = [HbCO], is 200–300 times that for O_2.

The TLV-TWA (ACGIH) is 25 ppm by volume. TLV stands for threshold limit value, TWA stands for time-weighted average, and ACGIH stands for American Conference of Governmental Industrial Hygienists. The inhalation of air containing 400 ppm by volume of CO will result in headache and discomfort within 2–3 h. Inhalation of air containing 4000 ppm by volume proves fatal in less than 1 h. Inhalations of air in which the concentration of CO is high can cause sudden collapse with little or no warning.

Measurements The Danish fire-brigade carried out measurements in the aforementioned space after the compartment had been ventilated for approximately 1.5 h. 80 ppm was measured in the upper part, whereas 20 ppm was measured in the lower part. The door giving access to the compartment was subsequently closed and the measurements were repeated after 36 h. 690 ppm was measured in the upper part, whereas 555 ppm was measured in the lower part. Also after 36 h, the fire-brigade measured the CO-concentration in a compartment adjacent to the aforementioned space. In that compartment, the CO concentration appeared to be higher than 2000 ppm, the maximum that could be measured with the equipment at hand. Further measurements were carried out after the ship had been unloaded. CO could be detected in neither the aforementioned compartment nor in the space adjacent to that compartment.

CO Formation Wood is a combustible material. The burning of wood is a chemical reaction between wood and oxygen (O_2) at elevated temperature. The reaction between wood and O_2 proceeds, albeit at a low rate, also at ambient temperature. If, at the reaction between wood and O_2, there is an abundant supply of O_2, the reaction product is carbon dioxide (CO_2). Otherwise, CO is also formed. Wood pellets are a bulk material and it means that about 55% by volume of the bulk material is wood and about 45% by volume of the bulk material is, at unconditioned storage, air. That means that the wood pellets expose, in $m^2 kg^{-1}$, a large area to air. As a result of chemical reactions between wood and oxygen, CO is formed. The separations between, on the one hand, the bow thruster space and an adjacent compartment and, on the other hand, the cargo-space are not airtight and thus CO could enter, by both diffusion and natural convection, the mentioned compartments. The measurements carried out by the Danish fire-brigade prove that this actually happened. The main part of the CO in the bow thruster space and an adjacent compartment probably entered them as follows. A cargo-space ventilation duct passes through the space in which the accident happened. There is a ventilation duct inspection hatch in that compartment. An inspection after the accident revealed that the hatch was not airtight.

A CO concentration of 500 ppm by volume in air does not imply a significant lowering of the O_2 concentration in air. Air normally contains approximately 21% by volume of O_2 and a CO concentration of 500 ppm by volume is 0.05% by volume. So, a considerable CO level (in view of its toxicity) hardly affects the O_2 level.

Safety Procedures

Before the Accident   A procedure for entering an enclosed space was applicable for the Lady Irina. So, the hazard was known. However, that procedure was not adhered to for the compartment in which the accident happened. A shortened procedure was applicable. That procedure implied that, some time before entering the space, the door had to be opened. Personnel was allowed to enter the compartment after the door had been open for 15–20 min. During the day, the door was left open. The space was entered frequently during the day. Applying the official procedure would have been a roundabout way. Measurements of gas concentrations were not carried out.

After the Accident   A new procedure for enclosed spaces in all ships of the fleet to which the Lady Irina belongs is now applicable. The new procedure defines a "First Entry." A "First Entry" is any entry when the enclosed space is closed. The first person entering the space must carry an instrument with which it is possible to measure the concentration of various gases in the compartment. The instrument gives an acoustic signal if the concentration of a dangerous gas is too high or the O_2 concentration is too low. The MSA Altair 4X Multigas Detector is a typical instrument. It can test for, e.g. LEL (lower explosion limit), O_2, CO, hydrogen sulfide (H_2S), sulfur dioxide (SO_2), and nitrogen dioxide (NO_2). Four of these measurements can be carried out simultaneously. At least three of these instruments have to be available aboard a ship. The instruments must be maintained and calibrated with calibration gas regularly. A board on the door giving access to the enclosed space outlines the new procedure.

Remarks

The detailed description of the accident indicates that CO is a dangerous gas indeed because it cannot be detected by the natural senses of the body. The crew of the Lady Irina was aware of the risks associated with the cargo of the ship. However, it did initially not occur to the crew that the CO concentration in the mentioned compartment was too high.

The Danish fire-brigade could get both the chief engineer and the first chief out of the bow thruster space. The ship's crew could not do this. These two facts give rise to the question whether it would be necessary to have, on board of the ship, means available to, if need be, hoist a human from, e.g. the bow thruster space. However, such provisions have not been included in the new approach.

7.3 INDUSTRY

Lethal accident during maintenance of a phosphorus furnace at Flushing in The Netherlands in 2009 [3].

Event

Two employees of Thermphos at Flushing in The Netherlands entered a phosphorus furnace during maintenance activities on May 15, 2009 without a compressed air breathing apparatus. The air in the furnace probably contained less than 10% by volume of oxygen and that probably caused their death.

Thermphos

Thermphos at Flushing was part of Thermphos International B.V.. The company went bankrupt in 2012. Competition by a company in Kazakhstan was an important aspect in the considerations concerning the bankruptcy. The company had activities in Europe, North- and South America, and Asia. The location at Flushing was the main seat of the company. The production of white phosphorus, *ortho*-phosphoric acid, and sodium tripolyphosphate (NTPP) were the main activities at Flushing. Thermphos International B.V.. employed approximately 1200 people. The workforce at Flushing was 450.

Phosphorus

Phosphorus is an element indicated by P. As an element, phosphorus exists in two major forms, i.e. white phosphorus and red phosphorus. Phosphorus is never found as a free element on the earth due to its high reactivity. Phosphorus-containing minerals are usually present in their fully oxidized state, that is, as inorganic phosphate rocks.

White phosphorus emits a faint glow upon exposure to air due to oxidation (reaction with oxygen) of the material. The term "phosphorescence," meaning glow after illumination, derives from this property of white phosphorus. It is, therefore, usually stored under water in which it is insoluble. Red phosphorus does not emit a faint glow upon exposure to air.

White phosphorus is metastable. It is yellow when it is slightly impure. Its melting point is 44.1 °C and its boiling point at atmospheric pressure is 280.5 °C. White phosphorus can be converted into red phosphorus.

White phosphorus burns to phosphorus pentoxide (P_2O_5) and diphosphorus tetroxide P_2O_4, depending on the amount of oxygen present. Finely divided white phosphorus particles are extremely reactive and ignite spontaneously in air.

White phosphorus is used for the manufacture of *ortho*-phosphoric acid (H_3PO_4) mainly. 50% of the H_3PO_4 from white phosphorus is used for the manufacture of phosphates for detergents, whereas also 50% is used for the production of phosphates for foodstuffs, pharmaceuticals, and animal feeds. It is also possible to manufacture H_3PO_4 directly by the reaction between sulfuric acid and phosphate rock. However, H_3PO_4 thus obtained is less pure than H_3PO_4 obtained from white phosphorus. The reason is that, during its production, white phosphorus leaves the phosphorus furnace as a vapor that is subsequently condensed in water. The white phosphorus thus produced is pure.

Phosphorus Production

Thermal phosphorus manufacture (in an electric furnace) starts nearly always from fluoroapatite ($Ca_5(PO_4)_3F$). The reaction equation is

$$4Ca_5(PO_4)_3F + 18SiO_2 + 30C \rightarrow 3P_4 + 30CO + 18CaSiO_3 + 2CaF_2$$

The second reactant is silica in the form of flint pebbles or gravel, whereas the third reagent is coke. The three raw materials are, in a particulate form, continuously added to the reactor. The chemical reaction is endothermic, which means that heat must be supplied to the reactor. The reaction temperature is in the range of 1400–1800 °C and the pressure is slightly higher than atmospheric pressure to avoid the ingress of air as it contains oxygen. Figures 7.2 and 7.3 illustrate the construction of an electric furnace in which white phosphorus was produced at Flushing and in which the accident occurred. The top view shows that the top of the reactor has the form of an equilateral triangle with rounded corners. There are three electrodes situated at the corners of a smaller equilateral triangle. Electric currents pass from the electrodes to the floor of the furnace and the reacting mass is heated by the heat development due to the electric resistance offered by the reacting mass. The triangular arrangement of the electrodes avoids the existence of live and dead phases. The feed lines are near the electrodes. It is important that the grain sizes of the three raw materials are the same. That results in virtually constant gas permeability of the reacting mass. The lines through which the gaseous reaction products white phosphorus and carbon monoxide leave the furnace are visible in the upper part of Figure 7.2. Figure 7.3 shows the large dimensions of the furnace used by Thermphos; its diameter is approximately 9 m. Large furnaces, like

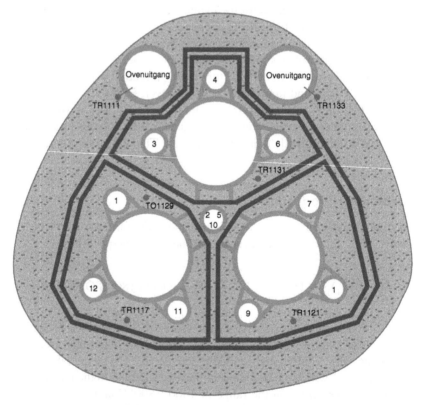

Figure 7.2 Top view of the phosphorus furnace. Ovenuitgang, Furnace exit.
Source: Courtesy of Dutch Safety Board, The Hague, The Netherlands.

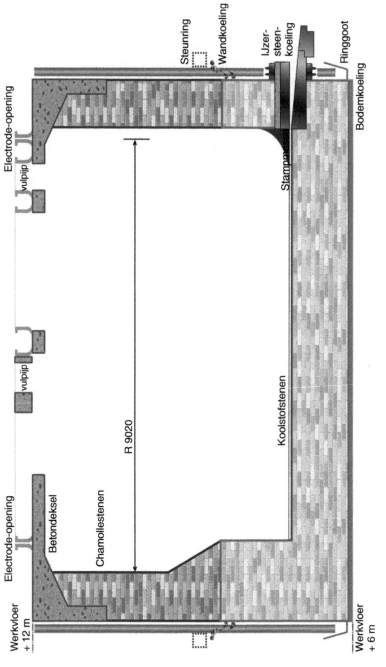

Figure 7.3 Cross section of the phosphorus furnace. Werkvloer, shop-floor; Electrode-opening, opening for electrode; vulpijp, filling pipe; Betondeksel, concrete lid; Chamollestenen, Chamolle stones; Steunring, support ring; Wandkoeling, wall cooling; Koolstofstenen, carbon stones; Stampmassa, pound mass; IJzersteenkoeling, cooling for iron stone; Ringgoot, annular drain; Bodemkoeling, bottom cooling. *Source:* Courtesy of Dutch Safety Board, The Hague, The Netherlands.

the furnaces used by Thermphos, have electrodes with a diameter of 1.3–1.5 m.

The other two reaction products leave the reactor as slags. First is the calcium silicate slag. Approximately eight metric tons of calcium silicate per metric ton of white phosphorus leave the reactor continuously as a viscous liquid. Its temperature is approximately 1350 °C. A second, much smaller, slag flow leaving the reactor intermittently is ferrophosphorus. Oxides of iron are impurities in fluoroapatite. They give rise to the formation of ferrophosphorus.

The gases leaving the reactor enter an electrostatic precipitator for the removal of dust. Next, the virtually dust-free gases, having a temperature of approximately 400 °C, enter a condensation system. Gaseous phosphorus is condensed in wash towers where it is brought into contact with water. Carbon monoxide passes the condensation step and is processed elsewhere.

Description of the Accident

The accident occurred in the early hours of May 15, 2009. Two phosphorus furnaces were being maintained at that time. Thermphos had three phosphorus furnaces in operation at Flushing. The maintenance activities at the furnace in which the accident occurred were almost ready. An operator got the request to hoist a faulty old valve into the furnace through an electrode opening. The probable reason for this request is that Thermphos wanted to get rid of the valve. The molten iron of the valve would add to the ferrophosphorus slag flow. The electrode opening mentioned was the only opening of the furnace as the two other electrodes had already been installed. He asked a colleague to assist him. The two employees noticed dust in the air in the empty phosphorus furnace. After the accident, it has been concluded on the basis of an investigation that the dust particles must have consisted of P_2O_5 or part of the dust particles must have consisted of that material. They decided to enter the furnace without breath protection because they could not find breath protection. The operator who had received the request to hoist the valve into the furnace climbed down a ladder and reached the floor of the furnace. The operator wanted to remove the valve from the tackle. He asked his colleague to assist him. The colleague then started to climb down the ladder. At that moment the operator in the furnace told him to leave the furnace because "something is wrong." The colleague thereupon left the furnace. The operator in the furnace also wanted to leave the furnace and started climbing the ladder. However, he fell 3 m down from the ladder on the bottom of the furnace. His colleague then informed the shift leader who worked in the vicinity. The shift leader asked for a compressed air breathing apparatus but did not wait for this provision and entered the furnace to provide first aid to the operator lying on the bottom of the furnace. He too fell 3 m down from the ladder while climbing down. Two Thermphos employees subsequently entered the furnace to assist their colleagues. These employees both wore a compressed air breathing apparatus. They reanimated their colleagues in the furnace and removed them from the furnace. Reanimation in the furnace and removal from the furnace lasted approximately 1 h. Reanimation was continued outside the furnace. The employees that had been lying in the furnace without breath protection died in the morning of May 15, 2009. The employee who had assisted his colleague for hoisting the valve and had been in the furnace for a short time was admitted to the hospital for observation.

The Cause of the Accident

Table 7.1 appears in [3]. All times in the table are times after the accident happened. The ppm values (parts per million) stated in the table are probably by volume. The height of the furnace is approximately 5 m. It is probable that a low oxygen concentration has caused the unconsciousness and the subsequent death of the two Thermphos employees. Humans faint almost immediately in air containing less than 10% by volume of oxygen. Prolonged exposure to such an atmosphere results in death.

The furnace was not ventilated at the time of the accident. It appeared that inert gas had leaked into the furnace via a raw material feeding line. During production, inert gas enters this raw material feeding line to prevent ingress of carbon monoxide. A valve in the inert gas line had not been fully closed.

Table 7.1 Concentrations of oxygen, phosphine, and carbon monoxide in the Thermphos Furnace after the accident.

Time on May 15, 2009 (h)	Oxygen (% by volume)	Phosphine (ppm)	Carbon monoxide (ppm)	Height[a] (m)
About 03.15	13.3	17.7	21	10
About 04.24[b]	5.5	33	62	7
About 04.24[c]	7.1	29	59	7
Note[d]	20.1	1		
About 11.30	20.9	0	0	

[a]Meters above ground level; see Figure 7.3.
[b]Instrument No. 1.
[c]Instrument No. 2.
[d]After activation of the main ventilation.

Carbon Monoxide, Phosphine, and Phosphorus Pentoxide

The question may be raised whether the two Thermphos employees could have been harmed as a result of the measured carbon monoxide and phosphine levels and the observed dust in the furnace.

First, the toxicity of carbon monoxide will be reviewed shortly [2]. See also Section 7.2. Carbon monoxide was in focus at Thermphos because it is a by-product of the reaction in which white phosphorus is produced. It follows from the data in Section 7.2 that carbon monoxide cannot have been the cause of the death of those two employees.

Next, the toxicity of phosphine will be mentioned [4]. Phosphine was in focus at Thermphos because it is known that electrodes resorb phosphine. Phosphine is a highly poisonous gas. Its formula is PH_3. For humans, concentrations of 400–840 mg m^{-3} over a period of 30–60 min are extremely dangerous or even lethal, whereas concentrations of 140–260 mg m^{-3} are tolerated under these conditions. However, one death has been observed following exposure to 10 mg m^{-3} for 6 h. To convert mg m^{-3} into ppm by volume at atmospheric pressure and 20 °C, multiply by 0.70. Also in view of the short time of exposure, it is improbable that phosphine has been the cause of the death of those two Thermphos employees.

Finally, the effect of the dusty atmosphere is discussed. As stated, it has been concluded on the basis of an investigation that the dust particles in the furnace at the time of the accident must have consisted of P_2O_5 or part of the particles must have consisted of that material. The Dutch Safety Board does not exclude the harmful effect of dust on the two affected Thermphos employees. I think that it is improbable that the dust has caused their death.

Remarks

Thermphos erred in not considering the furnace to be an enclosed space during the final phase of the maintenance activities. The furnace should, according to Dutch regulations, have been considered an enclosed space during that phase of the maintenance activities [5]. The definition of an enclosed space is, according to [5]: "A closed or partially open environment having or not having a narrowed access and insufficient or bad natural ventilation, which is not designed for the stay of humans, and where activities occur that entail risks concerning safety, health and well-being." The Dutch regulations require a detailed analysis of the risks of an enclosed space. Such an analysis could have brought to light that it would have been necessary to block the inert gas line to eliminate the possibility of inert gas flowing into the furnace. Probably Thermphos did not consider the possibility of inert gas flowing into the furnace. Such an analysis could also have led to the conclusion that it would have been necessary to have means available to, if need be, evacuate persons from the furnace.

7.4 SOCIETY

7.4.1 Fire in a Nightclub at West Warwick, Rhode Island in the United States in 2013

Event A fire occurred in The Station Nightclub at West Warwick, Rhode Island in the United States in the evening of February 20, 2003 [6]. A band performing that evening used pyrotechnics that ignited polyurethane foam lining the walls and the ceiling of the part of the building where they performed. The foam provided acoustic insulation. The fire spread quickly along the walls and the ceiling. Once most of the foam was consumed, the fire transitioned to a wood frame building fire. Hundred people lost their lives in the fire. The main cause of death was probably simple asphyxiation.

Additional Remarks The Station Nightclub was a single-story building having an area of approximately $412\,m^2$. The polyurethane foam present in The Station did most probably not contain fire retardants. It has been estimated that the conditions in the greater part of the nightclub would have led to severe incapacitation or death within approximately $1.5\,min$ after ignition of the foam for anyone remaining standing, and not much longer for those occupants close to the floor.

The capability to suppress the fire in its early stage of growth was insufficient primarily because automatic fire sprinklers were not installed. The building was equipped with hand-held fire extinguishers. The situation in 2003 was that sprinklers would have to be installed in new constructions; however, such provisions would not have been required for an existing structure like The Station Nightclub. The investigators of the fire recommend the installation of automatic sprinkler systems according to NFPA 13 for places such as The Station Nightclub. NFPA stands for National Fire Protection Agency (U.S.A.).

A heat detection/fire alarm system had been installed in The Station Nightclub, which activated (sound and light strobe) $41\,s$ after ignition of the polyurethane foam, by which time the crowd had already begun to move toward the exits.

Three exits were available: the double doors of the front main entrance on the north (limited by the single door into the vestibule), the single door on the west near the performance platform, and the single door on the east near the main bar. There were no emergency exits.

Egress from the nightclub was hampered by crowding at the main entrance to the building. The probable number of occupants was in the range 440–458, whereas the occupant limit amounted to 420 people. The building had several windows. These windows became the secondary route of escape.

7.4.2 Slurry Silo at Makkinga in The Netherlands in 2013

Event A lethal accident occurred at a livestock farm for dairy cattle at Makkinga in The Netherlands on June 19, 2013 [7]. An employee of a company specialized in slurry silo cleaning was cleaning a silo while wearing breath protection. The cleaning was necessary because the silo mixer had to be repaired. A second man working for the same company watched his colleague outside the silo standing on a ladder while the first man was working. The man inside the silo became unwell. The man outside the silo thereupon called for assistance and subsequently descended into the silo by climbing down a second ladder. Three further men working at the farm subsequently also descended into the silo. The stock farmer was one of these three men. One of these three men was, while he climbed down the ladder, instructed to go back because it was not safe in the silo. The three men having entered the silo to assist the man cleaning the silo also became unwell. The unwellness of all four men present in the silo has most probably been caused by the inhalation of air containing toxic gaseous components resorbed by the slurry in the silo. Those toxic components are hydrogen sulfide and hydrogen cyanide. Another person present at the farm called the emergency number. The fire-brigade succeeded in making a hole in the silo wall to evacuate the four men. Three men had by that time died in the accident and one was hospitalized heavily injured. The heavily injured man has partially recovered but will probably not recover completely.

The Farm and Procedures The farm is in the northern part of The Netherlands. At the time of the accident, there were approximately 110 dairy cattle at the farm. Both the faeces and the urine of the cattle in the stables were collected together in a cellar under the cattle. The cellar is called the slurry cellar. Typically, a slurry cellar has a depth of 2 m. The slurry is transferred to farmland again. In The Netherlands, it is not allowed to transfer slurry to farmland from about September 1 till about February 1. The reason for this standstill order is that nutrients can be taken up by vegetation only if the vegetation is present and if it is growing weather. The aim is to prevent pollution of ground-water, surface water, and air. Furthermore, it must be possible to store the slurry production of minimum 7 months in a silo. The rationale of the 7-month storage time instead of 5 months is weather and the availability of personnel to transfer the slurry to farmland. The livestock farm at Makkinga has a silo having a total volume of 924 m^3. Its diameter is 14 m and height is 6 m. Slurry is, typically once a month, transferred by means of a pump from the slurry cellar to the slurry silo. The slurry in the cellar is not homogeneous. It is necessary, to enable pumping, to regularly homogenize the contents of the slurry cellar. That process step is called mixing, and it is indeed performed by means of a mixer. The slurry, on arrival in the silo, segregates again. Before the transfer from the silo to a tankcar, it is often necessary to homogenize the slurry again by means of a mixer.

Gaseous Toxic Components

Enclosed Space and Mixing Two different situations can be considered. The first one is the situation when slurry is mixed in the cellar under the animals. This first situation considers what happens to people and animals present at ground level. Ground level refers to both inside and outside the stable. The second one is the situation where slurry is present in an enclosed space accessible to humans or animals. An enclosed space can be a cellar, a silo, a tank, or a container. The accident at Makkinga occurred in a silo.

Measurements The ppm values (parts per million) stated in this section are probably by volume. Measurements have been carried out at 22 dairy cattle livestock farms in the northern part of The Netherlands before and during mixing in (probably) 1986 [8]. The measuring period lasted approximately 1 year. Thirteen farms of the 22 had already experienced problems concerning toxic gases emerging from slurry, whereas 9 farms had not yet experienced such problems. The problems were that mainly cattle became unwell and that in some instances, died. Mixing occurred, on average, once per 3–4 weeks and lasted 2–4 h.

First, measurements were taken before mixing. On average, 0.055 vol.% of carbon dioxide (CO_2) and 3 ppm of ammonia (NH_3) were found in the stables of the 13 problem farms. In the stables of the 9 reference farms, 0.039 vol.% of CO_2 and 4.9 ppm of NH_3 were found. Measurements concerning other compounds were not carried out before mixing. The measurements were carried out in the stables at the height of an animal.

Table 7.2 Analytical data of air in dairy cattle livestock farms during mixing – averages of the highest values measured at the farms.

Compound	13 problem farms	9 reference farms
Hydrogen cyanide (HCN)	165 ppm	145 ppm
Hydrogen sulfide (H_2S)	376 ppm	339 ppm
Carbon dioxide (CO_2)	0.52% by volume	0.5% by volume
Ammonia (NH_3)	5.7 ppm	11.9 ppm
Methane (CH_4)	15.2% of the LEL[a]	10.2% of the LEL[a]
Oxygen (O_2)	20.63% by volume	20.70% by volume

[a] LEL stands for lower explosion limit.

Second, measurements concerning various compounds were carried out during mixing. They were started when the mixing was stationary. The measurements were carried out in critical areas. Area near the mixer and "stagnant" zones in a stable were defined as critical areas. The first series of measurements is summarized in Table 7.2. The figures in the table are averages of the highest values measured at the farms. The second series of measurements in both 13 problem farms and 9 reference farms concerns hydrogen cyanide only. The concentration of hydrogen cyanide was measured during mixing as a time weighted average (TWA) over a period of 15 min. The majority of the data is in the range 10–100 ppm. A significant difference between the two farm categories cannot be noticed. One measurement is as high as 144 ppm.

Toxicological Data
TLV stands for threshold limit value and TWA means time weighted average.

Hydrogen Cyanide TLV-TWA 10 mg m^{-3} (8.3 ppm) [9]. It is the highest concentration allowable for human exposure in air, 8 h a day, and 40 h per week. Exposure of humans to concentrations exceeding approximately 100 ppm can result in sudden collapse and stopping of breathing [10].

H_2S TLV-TWA 5–10 ppm by volume [11]. Exposure of humans to concentrations in air between 300 and 400 ppm for periods between 15 min and 1 h causes severe respiratory distress and acute asthenia. Exposure of humans to concentrations in air higher than 1000 ppm causes immediate loss of consciousness and respiratory distress.

CO_2 TLV-TWA 0.5% by volume [12]. Exposure of humans to air containing 5% by volume of CO_2 causes an increase in the breathing rate by a factor of approximately 3. Concentrations higher than 5% by volume rapidly cause unconsciousness and death.

NH₃ TLV-TWA 25 ppm [13].

Evaluation of the Measured Data   The concentrations of HCN and H_2S are high. The concentrations of CO_2 and NH_3 are harmless. A vapor/air explosion cannot occur. The air in the stables in question contains the right amount of oxygen. The accident at Makkinga in 2013 occurred in an enclosed space. The measurements in the northern part of The Netherlands in (probably) 1986 make it plausible that the Makkinga accident has been caused by HCN and/or H_2S.

Prologue of the Accident   Approximately $36\,m^3$ of purge water had been added to the silo on February 16, 2013. The purge water came from a pig livestock farm. It had been used to remove NH_3, dust, and smelling gaseous components from stable air in a scrubber. Purge water is added to the slurry because it contains nitrogen in the form of NH_3. Moreover, it decreases the viscosity of the slurry and thus facilitates the distribution on the farmland. The farmer transferred approximately $800\,m^3$ slurry from the silo on February 18, 2013. However, it was noticed during the distribution of the slurry on the farmland that the slurry had not been mixed properly. It was assessed that the silo mixer had not functioned adequately. Slurry remained in the silo. The slurry height in the silo was about 90 cm. The farmer then asked a specialized company to repair the mixer. It is necessary to empty the silo before the repair can be attempted.

Detailed Description of the Event   See Figure 7.4. Two employees of the company specialized in silo cleaning and repair arrived at the farm in the morning of June 19, 2013. They started the work at approximately 07.30 h while wearing watertight suits. A ladder was put up against the silo wall, a manhole in the roof of the silo was opened, and a second ladder was installed at the other side of the silo wall to be able to descend into the silo. One of the two employees entered the silo. He had equipped himself with breath protection. The breath protection consisted of a cap through which air was passed into the silo. The air came from a compressor located outside the silo. The compressor was driven by an electromotor. It was assessed after the accident that the compressor still could function. The air passed through a hose to the cap. The employee in the silo also wore detectors to detect dangerous concentrations of H_2S and low O_2-concentration in the air. The second employee stood, while the first employee worked, on the ladder outside the silo and watched the first employee. The work in the silo proceeded as follows. The employee in the silo forced the slurry toward a central outlet in the silo bottom by means of a water jet. The water came from a water container outside the silo. Next, the slurry was pumped into a

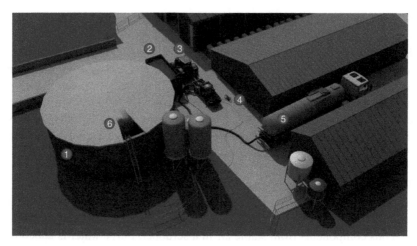

Figure 7.4 Approximate reproduction of the situation during the accident at Makkinga in 2013. (1) Slurry silo; (2) container filled with water; (3) tractor and car with pump; (4) breathing air compressor; (5) tankcar for slurry transfer; (6) opened manhole in silo roof. *Source:* Courtesy of Dutch Safety Board, The Hague, The Netherlands.

tankcar having a volume of $17\,m^3$. A filled tankcar was emptied on the farmland. A second tankcar having a volume of about $36\,m^3$ arrived at the farm at approximately 10.30 h to assist at the slurry removal. At that time, the driver of the 17-m^3 tankcar had left the farm to empty the tank on the farmland. The driver of the 36-m^3 tankcar then connected his tankcar to the silo and stayed at the farm.

The employee in the silo became unwell at about 11.00 h. The two employees of the specialized company had, by that time, worked for about 3 h. The height of the slurry layer had been decreased from about 90 cm to about 10 cm. The second employee watching the man in the silo called for help and then descended into the silo without breath protection. The driver of the 36-m^3 tankcar also went into the silo without breath protection. The farmer also descended into the silo without breath protection. He had been working on the farm with a gardener. The gardener also descended into the silo. However, when he had almost reached the bottom, the farmer told him to go back because it was not safe in the silo. The gardener succeeded in climbing the ladder and leaving the silo. He saw the farmer turning the victims so that they could breathe. The farmer then also tried to get out of the silo. He came as far as halfway the ladder and then fell back into the silo.

The farmer's father thereupon called the emergency number. The fire-brigade succeeded in making a hole in the silo wall and recovered the victims. However, three men had already died by that time and the driver of the 36-m^3 tankcar was heavily injured. The heavily injured man has recovered partially but will probably not recover completely.

The Compass of the Accident   The data and figures in this section apply for The Netherlands. There are about 20 000 dairy cattle livestock farms. There are more than 8000 pig livestock farms. The number of dairy cattle is approximately 1 500 000 and the number of pigs is about 7 000 000. Thus, the average number of dairy cattle per farm is about 75 and the average number of pigs per farm is about 875. The farm at Makkinga where the accident happened is slightly larger than the average farm.

The annual slurry production from cattle amounts to more than 65 000 000 metric t. About 50 000 000 metric t are annually transferred to the farmer's own farmland. The remaining 15 000 000 metric t per annum are used elsewhere, the greater part is transferred to a different farmland. Safety issues concerning slurry exist for slurry originating mainly from dairy cattle and pigs. More problems are experienced at dairy cattle livestock farms than at pig livestock farms.

Before 1987, silos did not have a roof. As from 1987, new silos should have a roof. As from January 1, 2013, all silos have to have a roof. The rationale of the gradual closing of silos is the reduction of emissions. Still, silos have an open connection with the atmosphere.

Low-emission grate floors have been installed in stables in the last couple of years. Low-emission grate floors allow the passage of manure from the stables through slits into manure cellars. However, the escape of gases from the manure cellars through the same slits into the stables has been made difficult by, e.g. rubber strips. At farms where the dairy cattle stays indoors, new stables and extensions have to be low-emission stables and low-emission extensions. As from 2015, that statement is valid for all new buildings and extensions. However, these measures are not sufficient to prevent HCN and H$_2$S entering the stables when mixing is performed.

Slurry in a dairy cattle livestock farm cellar tends to develop a hard and dry crust at the surface. The slurry must therefore be mixed regularly. The slurry is often also mixed in a silo before transfer to a tankcar. As a rule, slurry from a pig livestock is directly, i.e. without mixing, pumped into a silo. Such slurry is often mixed in a silo before the transfer to a tankcar.

In recent years, on average, three serious accidents occurred annually in The Netherlands. Almost 90% of the accidents occurred in a silo, cellar, or tank till approximately 5 years ago. More than half of the accidents in the last 5 years occurred in the stables or in the environment of the stables during mixing the slurry in the slurry cellars.

Addition of Materials to the Slurry   It has been stated previously that $36\,m^3$ of purge water had been added to the silo at Makkinga on February 11, 2013. The background has also been indicated. Transferring purge water to farmland is allowed in The Netherlands; however, mixing purge water with slurry is not allowed. The reason is that additions to slurry could mix up the bookkeeping of materials transferred to farmland. Additions of materials to slurry are not actively checked by the authorities in The Netherlands. The addition of further materials also occurs, e.g. the addition of fertilizers. It is possible that addition of materials to slurry has an effect on the formation of gaseous toxic gases.

Protections Used and Remarks

Breath Protection   The breath protection used by the employee working in the silo consisted of a cap on the head of the employee through which air was passed on to the silo. Thus, there was an open connection between the atmosphere in the cap and the atmosphere in the silo. Such protection is inadequate when there are high concentrations of dangerous substances in the air of the silo. It is neither suitable for spaces in which the air contains less than 17% by volume of oxygen. A compressed air breathing apparatus would have been required for the man working in the silo at Makkinga in June 2013. Moreover, a second compressed air breathing apparatus should have been present at the farm. The employee working in the silo wore detectors to detect a dangerous concentration of H_2S and too low a concentration of O_2. These detectors are useful but cannot replace breath protection. In addition, the accident has possibly rather been caused by HCN than by H_2S. As to the former compound, the employee working in the silo did not wear a detector to detect dangerous concentrations of HCN.

Construction   The rule to have all slurry silos equipped with a roof as from January 1, 2013 to protect the environment turns the silos into enclosed spaces. Thus, the risks for personnel working in the silos increase.

Tackle   Means to evacuate persons from the silo if need be were absent. Such means should have been present.

Hatches   There was no hatch in the silo wall at ground level that could have been opened to assist and evacuate persons from the silo if need be. I think that such a hatch should have been present.

Stairs   Putting up ladders at both sides of the silo wall to get into the silo is improvising. It would have been better to have a permanent stairs at the outside of the silo with a platform at the top of the silo. Both stairs and platform would have to be provided with hand-rails. A permanent ladder at the inside of the silo could be used for descending into the silo.

Ventilation   The presence of a hatch in the silo wall at ground level would enable forced ventilation through the silo by means of a fan. The air could leave at the top of the silo through the manhole.

Concluding Remarks   Four concluding remarks are made. The first one is that the safety measures during the cleaning of the slurry silo at Makkinga were inadequate. The second one is that the activities at both dairy cattle livestock farms and pig livestock farms have been scaled up in the last decades. At the same time, the rules have been tightened. The silos must have a roof and the period in which it is not allowed to transfer slurry to farmland has increased. The safety aspects have insufficiently been taken into account at the scaling up of the activities at dairy cattle livestock farms and pig livestock farms and the tightening of the rules. The third remark is that it is not good practice to add materials to slurry without knowing the effect it can have on the formation of toxic gases. The fourth and final remark concerns the standard of safety measures for the two types of livestock farms mentioned. The latter aspect receives the attention of the Dutch Safety Board. In the (petro)chemical industry, storage tanks that have contained chemical compounds are cleaned. This activity can be compared to the cleaning of a slurry silo for the two types of livestock farms mentioned. The Dutch Safety Board has asked a company specialized in cleaning industrial tanks to design a procedure for cleaning a slurry silo still containing a layer of viscous slurry. It strikes that this procedure takes better account of the risks of cleaning slurry tanks than the procedure in actual fact adhered to at Makkinga on June 19, 2013 [14].

REFERENCES

[1] Dutch Safety Board (2015). *Carbon Monoxide in Bow Thruster Space*, 1–22. The Hague, The Netherlands: Dutch Safety Board (in Dutch).

[2] (2001). Carbon Monoxide. Ullmann's Encyclopedia of Industrial Chemistry, Wiley Online Library.

[3] Dutch Safety Board (2013). *Lethal Accident during Maintenance of a Phosphorus Furnace*, 1–18. The Hague, The Netherlands: Dutch Safety Board (in Dutch).

[4] (2001). Inorganic Phosphorus Compounds. Ullmann's Encyclopedia of Industrial Chemistry, Wiley Online Library.

[5] SGS Nederland B.V. (2011). *Working Safely in Enclosed Spaces*, 1–42. The Hague, The Netherlands: Sdu Uitgevers bv (in Dutch).

[6] Grosshandler, W., Bryner, N., Madrzykowski, D., and Kuntz, K. (2005). *Report of the Technical Investigation of the Station Nightclub Fire*, vol. 1, xvii–xxvi. Washington, DC: United States Government Printing Office.

[7] Dutch Safety Board (2014). *Lethal Accident in a Slurry Silo at Makkinga*, 1–70. The Hague, The Netherlands: Dutch Safety Board (in Dutch).

[8] Factory Inspectorate and Animal Health Service, both in the Northern Part of The Netherlands (1988). *Dangers at the mixing of slurry in dairy cattle stables: the resorption of hydrogen sulphide and hydrogen cyanide* (in Dutch).

[9] (2001). Hydrogen Cyanide. Ullmann's Encyclopedia of Industrial Chemistry, Wiley Online Library.

[10] Braker, W., Mossman, L., *Effects of Exposure to Toxic Gases – First Aid and Medical Treatment*, Matheson Gas Products, East Rutherford, NJ, 1970, pp. 16, 17.

[11] (2001). Hydrogen Sulphide. Ullmann's Encyclopedia of Industrial Chemistry, Wiley Online Library.

[12] (2001). Carbon Dioxide. Ullmann's Encyclopedia of Industrial Chemistry, Wiley Online Library.

[13] (2001). Ammonia. Ullmann's Encyclopedia of Industrial Chemistry, Wiley Online Library.

[14] Dutch Safety Board (2014). *Lethal Accident in a Slurry Silo at Makkinga*, 30. The Hague, The Netherlands: Dutch Safety Board (in Dutch).

8

EXAMPLES FROM THE CHEMICAL INDUSTRY

8.1 INTRODUCTION

Five accidents that occurred in the chemical industry are described in Sections 8.2–8.6. The first case concerns failing of the cooling of a reactor. There was no back-up system for the cooling. The second case, described in Section 8.3, is the loss of control of a chemical reaction because the reaction temperature had risen to too high a value. Case 3, described in Section 8.4, concerns an explosion in a reactor, approximately 20 s later followed by an explosion in a separating vessel belonging to a second parallel reactor. An unforeseen chemical reaction occurred during the warming of catalyst pellets in the reactor in which the explosion took place. In the fourth case, described in Section 8.5, instructions were not adhered to and this caused a gas explosion in a furnace of a plant during the start-up of that furnace. The instructions were not adhered to because following the instructions was considered a roundabout and time-consuming way of starting the furnace. The fifth case concerns the explosion of a poultry feed additive in a closed contact dryer.

Safety in Design, First Edition. C.M. van 't Land.
© 2018 John Wiley & Sons, Inc. Published 2018 by John Wiley & Sons, Inc.

8.2 RUNAWAY REACTION AT T2 LABORATORIES AT JACKSONVILLE, FLORIDA IN THE UNITED STATES IN 2007

Event

A powerful explosion and subsequent fire occurred at T2 Laboratories, Inc., a chemical manufacturer at Jacksonville, Florida, on December 19, 2007 [1]. The explosion killed four employees of T2 and injured four employees of T2 and 28 members of the public, who were working in surrounding businesses. The explosion occurred at the production of the 175th batch of methylcyclopentadienyl manganese tricarbonyl (MCMT) in a 2450-gal (9.27-m^3) batch reactor. Following the incident, T2 has ceased its production operations.

Event Timeline

The process operator had an outside operator call the owners of T2 to report a cooling problem and requested they return to the site at 13.23 h. Upon their return, one of the two owners went to the control room to assist. The bursting disk broke, the reactor burst, and its contents exploded a few minutes later, at 13.33 h, killing the owner and the process operator and two outside operators who were exiting the reactor area.

Cause of the Accident

A runaway reaction occurred during the first step (the metalation step) of the MCMT process. A loss of sufficient cooling during the process likely resulted in a runaway reaction, leading to uncontrollable pressure and temperature rise in the reactor. The pressure rise in the reactor resulted in an explosion comparable to the explosion of 1400 pounds of TNT. The reactor contents, coming into contact with the atmosphere, ignited subsequently.

T2 Laboratories, Inc.

T2 was a small privately owned corporation that began operations in 1996. A chemical engineer and a chemist founded T2 as a solvent blending business and co-owned it until the incident. From 1996 to 2001, T2 operated from a warehouse located in a mixed-used industrial and residential area of downtown Jacksonville. T2 blended and sold printing-industry solvents; it also blended premanufactured MCMT to specified concentrations for Advanced Fuel Development Technologies, Inc., a third-party distributor.

In 2001, T2 leased a five-acre site in a North Jacksonville industrial area and began constructing an MCMT plant. In January 2004, T2 began producing MCMT in a batch reactor. MCMT production was the primary business operation by December, 2007. On the day of the accident, T2 employed 12 people.

MCMT

MCMT is an organomanganese compound used as an octane-increasing gasoline additive. Apart from blending, T2 manufactured MCMT themselves and sold it under the name Ecotane. Their process consisted of three steps that occurred sequentially within a single-batch reactor.

The Metalation Step of the 175th Batch

The manufacturing procedure used by T2 started with the charging of sodium, methylcyclopentadiene dimer, and dimethyl diglycol ether (purity 95% by weight) to the reactor. According to a recipe given by the Chemical Safety Board (CSB), the percentages by weight of these three compounds were, respectively, 10.59, 43.55, and 42.58. The first step is called the metalation step, because sodium, a metal, is one of the two reactants. The other reactant is methylcyclopentadiene dimer. Dimethyl diglycol ether was used as a solvent in this process. Sodium metal arrived packed in mineral oil to prevent oxidation and limit moisture contact with the metal. Some mineral oil, 3.28% by weight, was transferred into the reactor with the sodium. The mineral oil does not participate in or interfere with the reaction. The three aforementioned percentages by weight and 3.28% by weight add up to 100.00% by weight.

At the 175th batch, the operator began heating the batch to melt the sodium and initiate the metalation reaction at about 11.00 h on December 19, 2007. The metalation reaction is a reaction between sodium and cyclopentadiene. Cyclopentadiene dimer was added to the reactor. The name cyclopentadiene dimer indicates that two cyclopentadiene molecules have combined. First, the cyclopentadiene dimer was converted into cyclopentadiene monomer. That means that each combination of two molecules is split up into two single molecules. Next, cyclopentadiene reacted with sodium. Hydrogen gas was a by-product of this reaction. This gas was vented to the atmosphere. The pressure in the closed reactor was 53 psig (3.65 baro). Once sodium melted, at 210 °F (98.9 °C), the process operator likely started the agitator of the closed reactor. Heat from the reaction and the heating system (hot oil in a coil in the reactor) continued raising the reactor temperature. At a reaction temperature of about 300 °F (148.9 °C), the process operator likely turned off the heating system as specified in the manufacturing procedure, but heat from the reaction continued increasing the temperature of the reactor contents.

At a temperature of about 360 °F (182.2 °C), the process operator likely started cooling, as specified in the manufacturing procedure. A cooling jacket covered the lower three quarters of the reactor. City water was passed into the jacket at the bottom of the jacket and allowed to boil; steam from the boiling water vented to the atmosphere through an open pipe connected to the top of the jacket. Thus, the temperature of the water in the jacket rose to 100 °C. However, at the 175th batch, the temperature of the reactor contents continued to increase. The pressure increase caused by the temperature increase led to the explosion.

Chemistry of the Runaway Reaction

The CSB observed two exothermic reactions using the T2 recipe. The first reaction occurred at approximately 350 °F (176.6 °C) and was the desired reaction between sodium and MCPD. The second reaction was more energetic than the first one and occurred when the temperature exceeded 390 °F (198.8 °C). The second reaction was between sodium and the solvent dimethyl diglycol ether.

Remarks

A continuously running cooling system can be part of intrinsic process safeguarding as it does not need activation. The continuously running cooling system failed at T2 for unknown reasons. Extrinsic process safeguarding, i.e. a back-up system for the continuously running cooling system, was not in place. That is, redundancy was not practiced. See Chapter 1 for a discussion of the concepts of intrinsic continuous process safeguarding and extrinsic process safeguarding.

The temperature margin between the desired reaction and the undesired reaction is 22.2 K. That margin is acceptable.

It is likely that T2 management was not aware of the occurrence of an undesired reaction at 198.8 °C.

The reactor was equipped with a 4 in. bursting disk that appeared to be inadequate.

8.3 REACTIONS WITH EPOXIDES

The large-scale manufacture of these compounds has in practice given rise to incidents, e.g. at the reaction between epichlorohydrin and an N-substituted

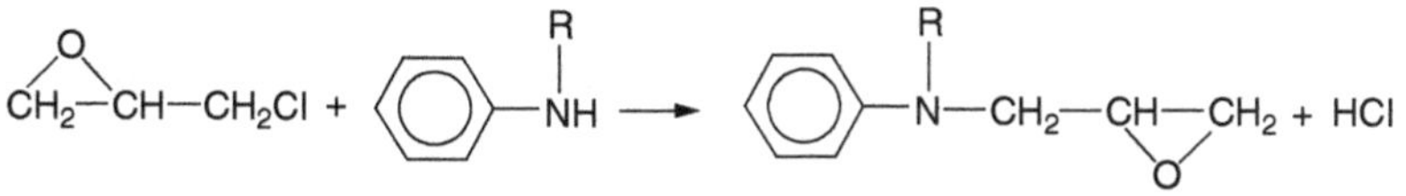

Figure 8.1 Equation of the chemical reaction between epichlorohydrin and an N-substituted aniline.

aniline [2]. The reaction equation is shown in Figure 8.1. This reaction was industrially carried out batchwise by adding the two reactants together and heating to 60 °C. At 60 °C, the reaction started and manually a switch was made from heating to cooling. One day, due to an operator's mistake, the temperature rose to 70 °C. The reaction could not be kept under control at 70 °C. The heat production rate exceeded the capability to transfer the heat to the cooling system. This shows that the rate of the reaction strongly increases with temperature. The temperature of the runaway reaction rose to 120 °C in 10 min and kept rising faster till an explosion followed. By that time the area had been evacuated. Personal damage did not occur.

A different reaction system had to be designed. The outcome of process research was to first add the full amount of N-substituted aniline to the reactor and to add epichlorohydrin in small portions next. Furthermore, the reaction temperature was raised. Each portion reacted immediately. The temperature of the reactor rose; however, the reactor contents could be cooled to the initial temperature by the cooling system. This procedure is in line with intrinsic continuous process safeguarding as described in Section 1.3. In addition, the reactor capacity increased due to the higher reaction rate.

8.4 EXPLOSIONS AT SHELL MOERDIJK AT MOERDIJK IN THE NETHERLANDS IN 2014

Event

Two explosions occurred at the Moerdijk Site of Shell at Moerdijk in The Netherlands at approximately 22.48 h on June 3, 2014 [3]. Both explosions occurred in the hydrogenation section of the SMPO2 plant. The first explosion concerned the bursting of Reactor No. 2 and the second explosion concerned the bursting of the separating vessel of Reactor No. 1. The exploding contents of the reactor and the separating vessel took fire. Two persons working in the vicinity of the plant were wounded. The explosions caused extensive damage.

The direct cause of the accident has been unambiguously established. Ethyl benzene (EB) reacted unforeseen with a catalyst. The possibility of the chemical reaction remained unnoticed and could at one time develop into a runaway reaction causing a pressure increase and the bursting of Reactor No. 2 and the separating vessel of Reactor No. 1.

Shell SMPO Plants

The acronym SMPO stands for Styrene Monomer Propene Oxide. The name SMPO2 refers to the second plant at the Moerdijk Site. Both styrene and propene oxide are products of an SMPO plant. Styrene is used to manufacture polystyrene, an insulating material. Propene oxide is used to manufacture propene glycol, a material used in food, cosmetics, and pharmaceutical products. The raw materials are EB, propene, hydrogen, and oxygen. The intermediate product methyl phenyl ketone (MPK) is converted into methyl phenyl carbinol (MPC) in the hydrogenation unit of an SMPO plant. This conversion is accomplished by the reaction between liquid MPK and hydrogen gas. The reaction occurs in the presence of a catalyst. A catalyst accelerates the rate of a chemical reaction without itself being changed chemically. The pressure at this process is rather high, e.g. 23 bar in the SMPO2 plant. The temperature at this process is also elevated. MPC, the product of the hydrogenation unit, is subsequently converted into styrene.

Shell exploits five SMPO plants worldwide. There are two plants at the Moerdijk Site, two plants at Seraya in Singapore, and one plant at Ninghai in China. The plants were built between 1979 and 2005.

The Hydrogenation Plant of SMPO2

The hydrogenation plant of SMPO2 is indicated as Unit 4800. The main components of this plant are two reactors, two separating vessels, one circulation pump, and one heat exchanger. The runaway reaction occurred in Reactor No. 2. The heart of the hydrogenation plant is the set of two reactors. They are continuous trickle-bed reactors. Figure 8.2 schematically depicts Reactor No. 2 and its separating vessel. A short description of how the plant is normally operating is given. MPK, the liquid to be converted into MPC, flows concurrently with hydrogen gas from the top to the bottom of the reactor. A distributor at the top of the reactor produces a shower out of the two-phase flow. The reactor has been filled with catalyst particles, the pellets. The liquid trickles down the catalyst bed. Each pellet is surrounded by a liquid film. The chemical reaction between the liquid and the

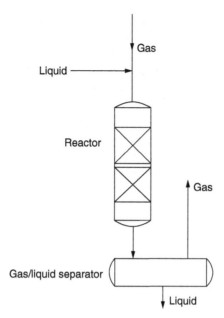

Figure 8.2 Continuous trickle-bed Reactor No. 2 with gas/liquid separator.

gas occurs at the surface of the catalyst. The diameter of Reactor No. 1 is 2.8 m and the diameter of Reactor No. 2 is 1.7 m. The height of the reactors is twice to thrice the diameter. The pellets are small cylinders having, typically, a diameter of 3.2 mm and a height of 3.4 mm. The two-phase flow leaving each reactor enters a horizontal separating vessel. The liquid collects at the bottom of the vessel, while the gas leaves the vessel at the top.

The Timeline and the Process Conditions of the Accident

The MSPO2 plant had been shut down on May 25, 2014 for maintenance. The main purpose of the stop was the replacement of the pellets in the two reactors. The new pellets were in place on June 3, 2014, and it was planned to start up Unit 4800. A step preceding the normal operation is the reduction of the catalyst. The catalyst contains among other compounds copper oxide, copper chromite, and copper chromate. These and other compounds contain oxygen. The reduction step comprises chemical reactions between these and other compounds on the one hand and hydrogen on the other hand. Hydrogen reacts with oxygen bound in the catalyst. The reduction was carried out by passing EB and hydrogen gas concurrently from the top of the reactors to the bottom of the reactors. Hydrogen reacted with the catalyst while EB did not. EB served as a carrier for the heat of the reduction reaction, and it was recycled and cooled in an indirect heat exchanger. The reduction reaction was carried out at 130 °C. Warming the pellets was a step preceding the reduction of the catalyst. This was carried out by passing EB and nitrogen gas concurrently from the top to the bottom of the reactors. The accident occurred during this warming step. EB served as a heat carrier to warm the pellets to 130 °C. Recycled EB was warmed in an indirect heat exchanger. The circulation with EB was started at 18.20 h. The EB flows to the two reactors were independent from each other. 250 kg h^{-1} of nitrogen gas passed through the two reactors in series and was subsequently vented. Reactor No. 1 was the first reactor entered by nitrogen. Warming of EB started at 20.15 h. The EB flow to Reactor No. 1 was stationary at 88 t h^{-1} between 20.15 and 22.48 h, the time of the explosions. The temperature of Reactor No. 1 increased gradually in this period and the three thermocouples along the height of the reactor indicated 120 °C at 22.48 h. The EB flow to Reactor No. 2 varied strongly between 2 and 30 t h^{-1} between 20.56 and 22.48 h. Compared to the warming of the pellets in Reactor No. 1, the warming of the pellets in Reactor No. 2 was delayed, e.g. the thermocouple at the bottom of the reactor indicated 60 °C at 22.48 h.

The Interruption of the Nitrogen Flow

The EB flow to both reactors had been started at 18.20 h. EB was collected in the separating vessels after having passed the reactors. The levels in the separating vessels varied strongly. The flow of nitrogen to the ventline was automatically closed at 22.15 h because of too high a level in the separating vessel of Reactor No. 2. The reason for this action is the necessity to avoid liquid being entrained into the ventline. Shortly after 22.15 h, the level in the separating vessel of Reactor No. 2 decreased. The operator could then have restored the nitrogen flow manually. However, the operator did not restore the nitrogen flow. Thus, as from 22.15 h, the nitrogen flow was reduced to zero. The system pressure rose to 7.8 bar, the nitrogen supply pressure.

The Cause of the Accident

Shell calculated that the EB flow to Reactor No. 2 should have been 16 t h^{-1} and the nitrogen flow 1700 kg h^{-1}. It is likely that the following scenario has caused the accident. There were dry zones in the catalyst bed of Reactor No. 2 during the warming of the pellets because of the varying EB flow and the small nitrogen flow. In these dry zones, there was EB on the surface of the catalyst but no flow of EB. A reaction occurred between EB and the catalyst in those dry zones. The heat of that reaction could not be carried away and, subsequently, hot spots developed in the bed. The temperature in those hot spots rose to 180 °C or to an even higher temperature. Reactions can occur at that temperature between, e.g. EB and copper oxide. The latter reactions developed into a runaway reaction leading to the destruction of Reactor No. 2 and the separating vessel of Reactor No. 1. The bottom temperature of Reactor No. 2 was 60 °C at 22.48 h, the time of the explosion. It is thus improbable that the explosion started in the bottom of Reactor No. 2. Likely, the explosion started in the upper part of Reactor No. 2. The high pressure caused by the explosion could more easily reach the separator of Reactor No. 1 than the separator of Reactor No. 2. There was a line between the top of Reactor No. 2 and the separator of Reactor No. 1. The relatively cold bed of Reactor No. 2 was between the upper part of Reactor No. 2 and its separator.

History of the Catalyst

The first catalyst bought by Shell is the Cu-1808T catalyst. This catalyst typically contains (% by weight):

Copper oxide	43
Chromic oxide	37.8
Sodium oxide	3.4
Silicon dioxide	10.8

It was supplied to the first plant, built in 1979 at the Moerdijk Site. According to the report of the Dutch Safety Board, it was established by means of experiments that the catalyst did not react with EB up to a temperature of 130 °C [4]. 130 °C is the temperature at which the catalyst is reduced. The first plant had a liquid-full hydrogenation reactor. Further plants were equipped with trickle-bed hydrogenation reactors. The catalyst Cu-1808T appeared to be not fully satisfactory for trickle-bed reactors. Moreover, the catalyst was expensive. Because of these two reasons, Shell tested three catalysts of different suppliers in the period 1999–2002. Regarding the comparison, the focus was on regular production. The G22-2 catalyst of a new supplier was selected as an alternative for the Cu-1808T catalyst. The catalyst G22-2 is a mixture of copper oxide, copper chromite, barium chromate, and silicon dioxide. It typically contains (% by weight):

Copper oxide	45–50
Barium compounds (as barium)	5–7

Shell labeled the new catalyst as a "drop-in" catalyst, i.e. neither plant equipment nor manufacturing procedures needed modification. The possibility of a chemical reaction between the G22-2 catalyst and EB was not checked.

Shell Experiments After the Accident

The chemical reactions between catalyst and EB were studied experimentally. Three different catalyst samples were warmed with EB and the evolution of carbon dioxide was checked. The evolution of carbon dioxide was considered a proof that, yes, chemical reactions between catalyst and EB had occurred. Reactions between catalyst Cu-1808T and EB started at 150 °C. Reactions between two versions of catalyst G22-2 and EB started at 100 °C. The first of these two versions concerned deliveries to Shell before and including 2010 and the second regarded deliveries to Shell after 2010. There was no significant difference between the latter two versions in the test.

Shell Explanation of the Experimental Results

The different behavior of catalysts Cu-1808T and G22-2 is caused by the difference in Cr(VI) content. Cr(VI) stands for six-valent chromium. Oxygen bound to Cr(VI) is more reactive vis-à-vis EB than oxygen bound to trivalent chromium. Cu-1808T contains 0.2–0.3% by weight of Cr(VI) and the two versions of G22-2 contain between 2.4% and 5.1% by weight (13 measurements of samples between 2004 and 2015). G22-2 tested between 1999 and 2002 contained 0.1–0.2% by weight of Cr(VI). The G22-2 deliveries to Shell started, at some point in time between 2002 and 2004, to contain substantially more Cr(VI). The Cr(VI) content of the catalyst G22-2 was not part of the specification. The supplier indicated the Cr(VI) content on the MSDS (Material Safety and Data Sheet). The MSDS was sent to Shell; however, the fact that the Cr(VI) content was higher than in 1999–2002 was not explicitly communicated to Shell by the supplier. The reason is that the Cr(VI) content was not part of the specification. Shell did not notice the higher Cr(VI) content.

Modifications Implemented by Shell After the Accident

The step of warming the pellets is now carried out by passing warm nitrogen gas through the catalyst beds. EB is no longer used to warm the catalyst.

The step of reducing the catalyst is now carried out by passing a mixture of hydrogen gas and nitrogen gas through the catalyst beds. EB is no longer used at the reduction step.

Remarks

According to the report of the Dutch Safety Board, Shell had concluded that EB was inert vis-à-vis the first catalyst, i.e. Cu-1808T, at temperatures up to 130 °C in 1977 [4]. EB did not react with Cu-1808T up to 130 °C. However, for such a conclusion, it would have been necessary to check the possibility of such a reaction at 150 °C. Such testing provides a safety margin of 20 K.

The catalyst manufacturer advised Shell in writing in 2010 and 2013 to reduce the catalyst in the gaseous phase. The catalyst manufacturer did not explicitly exclude other reduction procedures. Up to the accident in 2014, Shell reduced the catalyst in SMPO2 with a two-phase flow, i.e. a concurrent flow of hydrogen and EB from the top of the trickle-bed reactor to its bottom.

Working with a two-phase flow at both warming and reduction of the catalyst proceeds faster than working in the gaseous phase. However, the catalyst is replaced once in 3 or 4 years. Saving time here is relatively insignificant.

The accident happened during the warming step of the catalyst preceding the reduction of the catalyst. The catalyst is fit-for-use after the reduction step. The thought may come up whether it would be wise to ask the catalyst manufacturer to reduce the catalyst and to deliver a fit-for-use product. The reasoning here is that the catalyst manufacturer is knowledgeable concerning the manufacture of the catalyst, whereas the catalyst user is knowledgeable regarding the application. An aspect here is that the reduced catalyst reacts with oxygen in air. Thus, a reduced catalyst would have to be transported with precautions from the manufacturer to the user. The choice has therefore been made to reduce the catalyst in situ [5].

8.5   DSM MELAMINE PLANT EXPLOSION AT GELEEN IN THE NETHERLANDS IN 2003

Event

An accident occurred at the salt furnace of the second melamine plant (Melaf-2) of DSM Melamine Europe (DME) at Geleen in The Netherlands on April 1, 2003 [6]. A mixture of natural gas and air exploded in the furnace. The explosion lifted the roof of the vertical cylindrical furnace. Three employees of DME were standing on the roof, fell into the hot furnace, and lost their lives. Two further employees were slightly injured by the pressure wave. The furnace was damaged.

Stamicarbon Melamine Plants

Melamine (2,4,6-triamino-1,3,5-triazine) is produced from urea. It is used in the fabrication of melamine-formaldehyde resins for laminating and adhesive applications. Melamine is also used as cross-linker in heat-cured and high-solid paint systems.

There are several processes for the manufacture of melamine. DME uses the Stamicarbon process, which will be described shortly. The conversion of urea into melamine is carried out in the gas phase in a well-mixed continuous tank reactor at 7 bar and 400 °C with the aid of a silica-alumina catalyst. Ammonia and carbon dioxide are by-products of the chemical reaction. Urea is added to the lower part of the reactor as a melt having a temperature of 135 °C. The silica-alumina catalyst is fluidized in the reactor by gaseous ammonia having a temperature of 150 °C. Melamine is recovered from the reactor outlet gas by water quench and crystallization. The chemical reaction is endothermic, meaning that heat must be added to the reactor. Molten salt is circulated through heating coils in the reactor and provides the reaction heat. The molten salt thereby cools down and is reheated indirectly in a furnace by the combustion of gas. The accident happened at that furnace.

Melamine has a melting point of 350 °C. The material is, after the reaction, present in the plant as a powder. Hence, the plant parts after the reactor suffer from incrustations. These incrustations are removed by steam approximately once per fortnight. The cleaning procedure lasts approximately 12 h. Urea is not fed to the reactor when the reactor is being cleaned. Gaseous ammonia flows through the reactor. The reactor is kept hot during the cleaning step by circulating molten salt through the coil in the reactor.

The accident happened while the plant was being cleaned.

The Furnace of Melaf-2

See Figure 8.3. The furnace is a vertical cylindrical vessel having a height of 14 m and a diameter of 5 m. Natural gas or fuel-gas enter at the roof of the furnace and are burnt with air, which is also entering at the roof of the furnace. There is a coil in the furnace, which is close to the wall and through

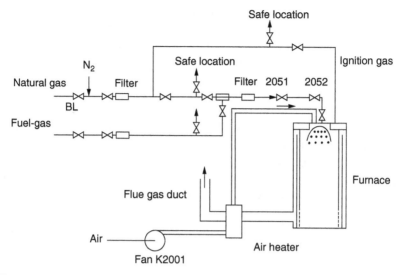

Figure 8.3 The salt furnace of a melamine plant. *Source:* Courtesy of Inspectie SZW of the Dutch Ministry of Social Affairs and Employment, Utrecht, The Netherlands.

which the molten salt circulates. The salt is heated by the gas flame while flowing through the coil and is then returned to the reactor where it provides the heat needed for the chemical reaction indirectly. The furnace was started up for the first time in 1999.

The salt circulating through the system was a mixture of potassium nitrate (55%) and sodium nitrite (45%). The salt had, at the time of explosion, a temperature of approximately 350 °C, the melting point of melamine.

The furnace load is low during the cleaning step. The cleaning step also provides the possibility to maintain the plant. Two maintenance activities were planned concerning the furnace. Two out of three filters should be cleaned and a leakage in the furnace roof should be repaired.

The two control valves 2051 and 2052 (see Figure 8.3) have the same function. This design had been chosen because of process safety reasons. If one control valve fails to function and stays in the open position, the other control valve is still active and performs the process control task. We see an example of redundancy.

Description of the Accident

The line related to natural gas should be considered only. The line related to fuel-gas did not play a role in the accident. The accident is related to the cleaning of the filters. The upper two filters in Figure 8.3 had to be cleaned. The cleaning procedure begins with the closing of the valve marked BL (Battery Limit). Natural gas cannot flow to the furnace anymore. First, the procedure for the left filter will be treated. By opening and closing valves, the gas in the line can be displaced by admitting nitrogen and leading the gas to a safe location. This design is called a design with block and bleeder. It is then possible to remove the filter, clean it, and reinstall it. Next, the procedure for the right filter will be discussed. The design for the right filter is not a design with block and bleeder. The control valves 2051 and 2052 were manually locked in the open position to enable the displacement of the gas with nitrogen into the furnace. It is then possible to remove the filter, clean it, and reinstall it. It was allowed to lock the control valves 2051 and 2052 in the open position for the displacement of the gas with nitrogen.

The line was filled with nitrogen after the reinstallment of the two filters. The nitrogen was removed by admitting natural gas through the manually controlled valve marked BL. Subsequently, the gaseous mixture was passed on into the furnace. The valves 2051 and 2052 were, during this admission of natural gas, still locked in the open position. Fan K2001 was started manually and locally shortly after the removal of nitrogen by the admission of natural gas. The explosion occurred a few seconds after the start of the fan. It was not allowed to lock the two control valves in the open position while admitting natural gas to the furnace. Instead, the two valves should, during the restart of the furnace, no longer have been controlled manually but by the process control system. The reason that the operators kept the two valves under manual control is that it was, when the two valves were controlled by the process control system, difficult to ignite the gas to restart the furnace because it was initially mixed with nitrogen. The ignition had to be activated several times and each time the ignition failed the furnace had to be flushed with air to displace inflammable material from the furnace. This was considered a roundabout and time-consuming way of starting the furnace.

The roof of the furnace having a diameter of 3.80 m was lifted by the explosion and pushed against the roof of the structure in which the furnace was installed. It fell back on the furnace and then made an angle of approximately 45° with the horizontal.

A Closer Look at the Explosion

About 26 nm^3 of natural gas had been admitted to the furnace. This corresponds with 59 m^3 at 345 °C and atmospheric pressure. The furnace volume was approximately 250 m^3. The natural gas concentration in the furnace would hence have been approximately 24% by volume if the gaseous mixture in the furnace would have been well mixed. The upper and lower explosion limits (UEL and LEL) are, respectively, 19.9% and 3.7% by volume at 345 °C and atmospheric pressure. This means that the natural gas content of the gas in the furnace would be somewhat greater than the UEL if the furnace contents would have been well mixed. It also means that the natural gas content of the gas in parts of the furnace could have been in between the UEL and the LEL if the gaseous mixture in the furnace would not have been well mixed.

Fan K2001 was started shortly after the natural gas entered the furnace. The specific mass of hot air in the furnace is 0.44 times the specific mass of air at ambient temperature. The specific mass of natural gas at ambient temperature is 0.55 times the specific mass of air at ambient temperature. Thus, the relatively cold natural gas collected at the bottom of the furnace. An oxygen content of 0% in the flue gas duct has actually been measured, proving that air in the flue gas duct was replaced by natural gas. The natural gas bubble at the bottom of the furnace was heated by the coils in the furnace and acquired buoyancy. The natural gas/air mixture moved upward and collected against the hot roof. Air was mixed with this natural gas/air mixture in a short time when fan K2001 was started. This caused a natural gas concentration in the gaseous mixture or part of the gaseous mixture between the UEL and LEL. It has been estimated that the temperature of the ceramic burner tip was between 800 and 1000 °C. Ignition at the burner tip, possibly in combination with self-ignition in the vicinity of the burner tip, is considered the most probable mechanism.

It has been estimated that the maximum pressure in the furnace as a result of the explosion was between 0.6 and 1.5 bar gauge.

The Role of Fan K2001

Fan K2001 was activated shortly after the nitrogen in the natural gas line had been displaced by natural gas. The explosion occurred a few seconds after fan K2001 had been activated. Normally, the fan was kept running when the filters were being cleaned. The fan did not run during the cleaning of the filters on April 1, 2003, because, simultaneously with the cleaning of the filters, the roof was repaired. The starting of the fan will have enhanced the mixing of the gas in the furnace.

Background Information Concerning the Explosion

Three gas filters can be distinguished in Figure 8.3. It was envisaged in the design stage that the three filters would rarely have to be cleaned. Because of the use of fuel-gas, two out of three filters had to be cleaned more often than expected. The filters had to be cleaned every 2 months. Due to this circumstance, the prescribed procedure for start-up had not been followed on April 1, 2003.

Modifications Implemented by DSM After the Accident

It was decided to stop using fuel-gas in the furnace of Melaf-2. This considerably reduced the cleaning frequency of the filters. The line-up of the filter after the mixer was changed. A design of block and bleeder was implemented. Nitrogen in the line can now be displaced to a safe location. A parallel spare filter for the filter after the mixer was installed. Finally, a physical coupling between valve BL and the process control system was made. This physical coupling makes it impossible to lock valves 2051 and 2052 in the open position when gas admission is possible. Several extrinsic safety measures were implemented additionally. For example, the gas supply is closed when fan K2001 is not running.

Remarks

The heart of the matter in this case is that instructions were not followed. That made it possible that natural gas entered the hot furnace while the furnace was not operating. A relatively unsatisfactory design seduced the staff and the operators to not adhere to the procedures. The fact that fan K2001 was not running at the time of the admission of the gas made it worse.

The roof of the furnace could have been a no-go area when the furnace is functioning. Functioning includes start-up.

8.6 DRYER EXPLOSION IN A DOW PLANT AT KING'S LYNN, NORFOLK IN THE UNITED KINGDOM IN 1976

Event

An explosion occurred at the Dow Company's chemical plant at King's Lynn, Norfolk, Great Britain, at approximately 17.10 h on June 27, 1976 [7]. The explosion caused the death of one Dow employee and extensive damage to the plant and buildings on the site.

The explosion concerned approximately 1300 kg of Zoalene, a poultry feed additive, which had been left inside a closed contact dryer for a period of 27 h after the drying operation had been completed. According to the report [7], the explosion was a detonation. The Zoalene had a temperature between 120 °C and 130 °C at the end of the drying operation. Under these circumstances, the product began to decompose with the evolution of heat. The evolution of heat per unit of time exceeded the heat loss to the surroundings per unit of time. This caused a temperature increase of the product. The decomposition rate increased due to the temperature increase. The runaway reaction ultimately led to detonation. The energy released in this event was approximately equivalent to the energy released by 200–300 lb (90.8–136.2 kg) of TNT (trinitrotoluene) on detonation.

The Product

The chemical name of Zoalene is 3,5-dinitro-*o*-toluamide, an aromatic compound. The material is used to prevent coccidiosis infections of poultry. The purity of the material is minimum 98% by weight. The material is very slightly soluble in water.

The Equipment

The product had been dried batchwise in a contact dryer. The moisture removed from the material by the drying operation was water. The dryer was a double-coned glass-lined steel dryer equipped with a jacket. The dryer was located at the ground floor of a building. During the drying operation, steam was supplied to the jacket, whereas vacuum was maintained in the dryer. Evaporated water could be condensed overhead and collected in a catchpot. The dryer rotated about its horizontal axis while drying was taking place. The rotational speed could be varied between 1 and 11 rpm. The dryer was charged manually and discharged by means of a conveyor. The diameter of the dryer was 1.8 m and the height was 2.34 m. See Figure 8.4.

The Process

The specific Zoalene drying operation was the tenth drying operation of a campaign to rework off-spec material. Dow decided to redry between 75 and 92 metric tons of this material. The purity was 96–98% by weight and it could be brought on-spec by redrying. Wall incrustations formed during the drying operations. To remove incrustations from the walls of the dryer, water had been added. The dryer was half filled with water. The dryer was then rotated and heated until the sides were clean. Such a cleaning operation was carried out at one out of approximately three drying operations. More Zoalene was added to make up a batch of approximately 1300 kg of

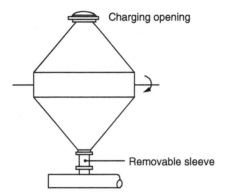

Charging opening

Removable sleeve

Figure 8.4 Double-coned contact dryer.

the product. The drying started at 15.00 h on June 25. The dryer was shut down at 14.00 h on June 26. The steam to the jacket was turned off and the vacuum released. The dryer was left with the charging opening lid clamped loosely in place. The explosion occurred at approximately 17.00 h on June 27, which means that the material had resided in the dryer after the drying operation for 27 h. At other drying operations, the batch was cooled by passing cold water through the jacket for about 40 min. The batch was normally cooled in order to permit immediate handling. However, in this case, cooling did not take place.

The Process Conditions

Steam having a pressure of about 40 psig (2.7 barg) was admitted to the jacket, resulting in a jacket temperature of 130–140 °C. A vacuum of about 25 in. Hg (97.7 mbar absolute pressure) was applied to the dryer. The water boiling point at this pressure is 45.5 °C. The rotational speed of the dryer was increased stepwise until it reached, after 4 h, 10 rpm. The water flow to the catchpot was an indication of the progress of drying. The endpoint was reached when the water flow came to a halt.

Product Safety Characteristics

After the accident, Dow established that typical production samples of Zoalene would self-heat to decomposition if held under near adiabatic conditions at 120–125 °C for 24 h.

The Royal Armaments Research and Development Establishment (RARDE) at Woolwich, Great Britain, established the explosive potential associated with Zoalene.

Remarks

Explosive properties of an aromatic compound having two nitro groups in the molecule could have been surmised.

The material should not have been left for 27 h in a closed process vessel at an elevated temperature.

As a general rule, reworking off-spec material entails more risks than regular production.

REFERENCES

[1] U.S.A. Chemical Safety and Hazard Investigation Board (2009). *T2 Laboratories, Inc. – Runaway Reaction (four killed, 28 injured), Report No. 2008-3-1-FL,* U.S.A. Chemical Safety and Hazard Investigation Board, Washington, DC, pp. i–viii, 1–69.

[2] Schierwater, F.W. (1971). The safe operation of exothermic reactions, especially in the liquid phase. *Industrial Chemical Engineering Symposium Series* *34*: 47–48.

[3] Dutch Safety Board (2015). *Explosions MSPO2 Shell Moerdijk*, 1–217. The Hague, The Netherlands: Dutch Safety Board (in Dutch).

[4] Dutch Safety Board (2015). *Explosions MSPO2 Shell Moerdijk*, 6. The Hague, The Netherlands: Dutch Safety Board (in Dutch).

[5] Shell Nederland Chemie B.V. (2015). E-mail titled MSPO2 and dated October 15, 2015.

[6] Factory Inspectorate of the Province Limburg in The Netherlands (2003) *Report of Findings, Administrative Investigation of the Accident in the DSM-Melamine Plant 2 at Geleen on April 1, 2003,* Factory Inspectorate, Directorate Major Hazard Control, Province Limburg, Geleen and Sittard, The Netherlands, pp. 1–95 (in Dutch).

[7] Health and Safety Executive (1977). *The Explosion at the Dow Chemical Factory, King's Lynn*, 1–19. London, Great Britain: Her Majesty's Stationary Office.

9

GAS EXPLOSIONS

9.1 INTRODUCTION

Two types of gas explosions are treated in this chapter. The first category includes gases formed from flashing inflammable liquids when their container collapses. Flashing liquids are liquids that, when stored under pressure, start to evaporate spontaneously when exposed to the atmosphere. The explosions of this category can be particularly violent because large quantities of material can be mixed with air in a short time. A well-known type of explosion in this category is the BLEVE (boiling liquid expanding vapor explosion). Sections 9.2–9.7 deal with flashing inflammable liquids. The second category concerns the leaking of an inflammable gas into the atmosphere or into an enclosed space. That can occur when, e.g. a natural gas line is damaged. A substantial gas flow into the atmosphere can be dangerous. Even a relatively small gas flow leaking into an enclosed space can be dangerous because it is possible to obtain an explosive gaseous mixture. An example of the latter type of accident is discussed in Section 9.8.

Generally, accidents caused by flashing inflammable liquids are more serious than those by gas leakages.

Safety in Design, First Edition. C.M. van 't Land.
© 2018 John Wiley & Sons, Inc. Published 2018 by John Wiley & Sons, Inc.

9.2   FLASHING INFLAMMABLE LIQUIDS

The accidents described in Sections 9.3–9.7 occurred with flashing inflammable liquids. Those liquids are particularly hazardous materials, and this aspect is dealt with in this section. An accident in which such a liquid was involved is described in Section 1.4. The liquid is LPG in Sections 9.3, 9.4, and 9.6. LPG stands for liquid petroleum gas. The liquid is propylene in Section 9.5 and 1,3-butadiene in Section 9.7. From a safety point of view, the transport, storage, and use of flashing inflammable liquids introduced a new dimension, compared to the transport, storage, and use of nonflashing inflammable liquids. Examples of the nonflashing inflammable liquids are petrol and diesel fuel. Propylene will be discussed as a typical example of a flashing inflammable liquid. The physical properties of propylene will be used for a calculation. The boiling point of propylene is −47.70 °C at atmospheric pressure. That means that propylene can only be kept in the liquid phase at ambient temperature under pressure. The pressure is dependent on the ambient temperature and will be in the range of 5–10 bara. When liquid propylene is ejected from a damaged vessel into the atmosphere, it is not in equilibrium with the atmosphere having atmospheric pressure. It can become in equilibrium therewith by cooling down to −47.70 °C. Its saturated vapor pressure is then equal to the atmospheric pressure. This cooling down is caused by rapid evaporation of a fraction of the propylene mass. That fraction will be estimated. It is assumed that the evaporation proceeds adiabatically, i.e. heat is not exchanged with the surroundings. The heat of evaporation of propylene at −47.70 °C is 437.49 kJ kg⁻¹.

The specific heat (heat capacity) of liquid propylene at −50 °C is 2.08 kJ kg⁻¹ K⁻¹.

One kilogram of liquid propylene, on flashing, cools down from ambient temperature to −47.70 °C. That could mean a temperature drop of about 75 K. The sensible heat made available thereby is $1 \cdot 75 \cdot 2.08 = 156.0$ kJ. That heat is used to evaporate propylene. The amount of evaporated propylene is

$$\frac{156.0}{437.49} = 0.357 \text{ kg} .$$

Thus, the fraction of propylene evaporated is 0.357. In actual fact, the fraction will be higher because the evaporation does not proceed adiabatically. Liquid propylene on the ground will be warmed by the ground.

So, more than one third of liquid propylene evaporates almost instantaneously when liquid propylene is, at ambient temperature, exposed to the atmosphere. A further aspect is the entrainment of droplets enhancing a BLEVE's effect. They are ignited like inflammable dust particles at a dust explosion. The specific mass of gaseous propylene exceeds the air specific mass by a factor of 1.45. This statement is valid when air and gaseous propylene have the same temperature and pressure. Thus, in that case, gaseous propylene spreads over the ground while mixing with air.

The discussion will now be continued by considering flashing inflammable liquids in general. Bursting or ripping up of a vessel containing such a material may be caused by heat or mechanical impact.

First, the effect of heat is considered. When the vessel is exposed to a fire, the saturated vapor pressure of the liquid in the vessel rises and thus the vessel pressure. Simultaneously, the heating of the hull material causes a loss of its strength. A combination of both effects may cause bursting of the vessel, and this results in a sudden availability of a large amount of flashing inflammable material. A BLEVE then occurs when the vapor/air mixture is ignited. A BLEVE caused by a fire is called a warm BLEVE. The BLEVEs at Mexico City, see Section 9.3, and Nijmegen, see Section 9.4, were caused by fires and were thus warm BLEVEs. The accident at Mexico City concerned a storage and distribution center, whereas the accident at Nijmegen concerned a tank-lorry.

Second, the effect of mechanical impact is considered. A distinction is made between a large damage of a vessel and a relatively small damage. In the case of a major damage, the vessel rips up and a BLEVE results. A BLEVE caused by mechanical impact is called a cold BLEVE. The BLEVE at Los Alfaques, see Section 9.5, was caused by the collision of a tank-lorry and was thus a cold BLEVE. In the case of a relatively small damage or leakage, one or more vapor/air explosions may result when vapor/air mixtures are ignited by, e.g. a flame or a spark. That occurred at Viareggio, see Section 9.6. A hole of 15 cm in size was punched into the hull of a wagon containing LPG.

A flashing inflammable liquid was involved at the incident described in Section 9.7. A wagon containing 1,3-butadiene was damaged; however, the liquid did not escape. We see a narrow escape.

A warm BLEVE is more serious than a cold BLEVE because, due to the heating of the flashing inflammable liquid, more energy is available.

A BLEVE produces a fireball. The duration of a fireball produced by a BLEVE depends on the amount of material involved and is typically in the range of 10–20 s [1]. A safe distance concerning a BLEVE of a tank-lorry could be 400 m. These figures are empirical values, and theoretical calculations are very complex. A loaded tank-lorry typically contains 30 m^3 of liquid.

9.3 MEXICO CITY IN 1984

Event

The accident occurred in a large LPG storage and distribution center at San Juan Ixhuatepec, 20 km north of Mexico City in the early morning of November 19, 1984 [2–4]. The facilities were owned by the Pemex (Petróleos Mexicanos) State Oil Company and consisted of six spherical tanks (two with a volume of 2400 m³ and four of 1600 m³) and 48 horizontal cylindrical tanks of different sizes. The total capacity of the storage tanks was 16 000 m³. They contained 11 000–12 000 m³ of LPG, i.e. a mixture of propane and butane, prior to the accident. The accident started with a leakage of LPG. A cloud formed, an ignition took place and caused a vapor/air explosion and a fire. The explosion and the fire caused further damage at the site, and between 05.44 and 07.01 h nine explosions occurred. One of these explosions was identified as a warm BLEVE. The accident took the lives of five Pemex employees and injured two employees. In the neighborhood close to the site, 550 people died and approximately 7000 people were injured.

The LPG Storage and Distribution Center

See Figure 9.1 and Table 9.1. The storage and distribution center was built in 1962. The two large spheres were added in 1981. The storage location was divided into separate sections by means of concrete walls of about 1 m high. The supply of LPG took place by pipelines and the shipping was by pipelines, tank-lorries, and wagons mainly. The site area was approximately 80 m long and 80 m wide.

LPG: A Flashing Inflammable Liquid

LPG is a flashing inflammable liquid and a mixture of propane and butane. The percentages by weight of these two liquids in LPG vary. The boiling points at atmospheric pressure of propane, i-butane, and n-butane are −42, −11.7 and −0.5 °C, respectively. It can only be kept in the liquid phase at ambient temperature under pressure. The pressure is dependent on the ambient temperature and a typical pressure is 6 barg. All storage vessels at San Ixhuatepec were, at the top, equipped with relief valves. The pressure at which a relief valve on a spherical vessel at this location opened was 150 psig (10.2 barg). LPG leaving uncontrolled a damaged vessel in which it was kept is not in equilibrium with the atmosphere, having the atmospheric

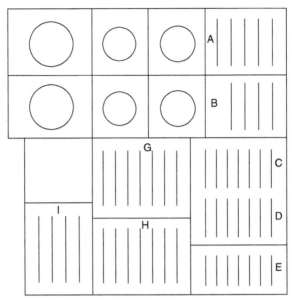

Figure 9.1 Tankfarm of the Pemex LPG installation. *Source:* Courtesy of Gelling Publishing, Nieuwerkerk aan den IJssel, The Netherlands.

pressure. It can become in equilibrium with the atmosphere by cooling down to approximately −20 °C. This cooling down is accomplished by rapid evaporation of a substantial fraction of LPG. This fraction could be about 0.25 for LPG. See Section 9.2 where, for propylene, the fraction is estimated. Gaseous LPG is 1.5–2 times heavier than air and an LPG-cloud spreads over the ground. In this stage, it does not yet mix properly with air. A fire is ignited when the spreading cloud meets an ignition source. An ignition source can be a flare or a running diesel motor. The fire then travels back to the point of release. The next phase is a vapor/air explosion because, by now, gaseous LPG and air are mixed at the point of release. The LEL

Table 9.1 Cylindrical storage vessels of the Pemex LPG installation.

Number of vessels	Volume per vessel (m³)	Section[a]
4	270	I
14	180	G, H
6	54	A, B
3	45	A, B
21	36	C, D, E

[a] See Figure 9.1.

(lower explosion limit) of propane is approximately 2% by volume in air. Next, the unvaporized portion of the LPG burns for, e.g. 20 min as a pool fire, and a huge conflagration sweeps the surroundings of the point of release. There is no crater, but at distances up to 100 m from the release point, pipelines are bent or torn from vessels due to the aforementioned explosion. The fire may lead to the collapse of storage vessels and BLEVEs may occur.

Detailed Description of the Event

November 19, 1984 was a Monday. The storage vessels had been filled during the weekend preceding November 19. The filling activities had been continued in the morning of that day. It is possible that a flange at one of the cylindrical vessels broke during the filling operation and that this failure caused the initial leakage. The ambient temperature was 7 °C at the time of the accident. There was a light north–east wind having a velocity of 0.4 m s^{-1}. A ground-level flare was burning all the time during the filling of the tanks in a device submerged in the ground for the burning off of excess gas. The flare was burning below the ground level, instead, as is usual, high above the ground. The strong winds prevailing locally could easily extinguish a flame above ground. The cloud formed by the leakage drifted toward a residential area and was possibly ignited by the ground-level flare. The fire traveled back to the point of release and caused a vapor/air explosion. The first explosion was followed by a huge fire. The fire and the explosion damaged the tanks at the site. Some tanks could no longer keep the LPG contained. The second explosion was a violent one and is characterized as a warm BLEVE [5]. It is possible that, at the second explosion, two smaller spherical tanks burst due to a fire. The fire caused a pressure rise inside the spheres and also caused a loss of strength of the sphere material. More explosions followed. It is possible that the other two smaller spherical tanks fragmented prior to a BLEVE.

Relief valves on storage vessels probably have functioned at San Juan Ixhuatepec. Relieved material will have taken fire. However, the relief valves could not cope with the pressure increase due to the heat input by the fire.

Casualties and Damage

The reason for the large number of fatalities and injuries is that the built-up area, with a high population density, was situated close to the site. When the plant was erected in 1962, the distance from the habitation was about

300 m. Under pressure from the large number of people moving in, the authorities had been unable to prevent a relatively primitive settlement from pushing ahead toward the depots. The shortest distance from the rows of houses to the storage tanks was reduced to 130 m. The majority of casualties occurred within the residential area close to the site, which reached out to roughly a distance of 300 m from the center of the storage location.

First, the damage at the site is discussed. See Figure 9.2. The two large spheres were not fragmented. Their supporting legs had collapsed due to fires. Both large spheres were damaged at the top and the contents had burnt away through the holes at the top. All four smaller spheres fragmented and some parts of these spheres had traveled about 400 m.

The cylindrical vessels had not been attached to their supporting structures. The vessels of Section I were relatively undamaged. Several vessels in Section G were displaced in a longitudinal direction. These vessels had kept their original order. Three vessels in Section H were, probably simultaneously, destroyed by fragmentation. The pressure wave of the explosion causing fragmentation displaced the mentioned vessels in Section G. The remaining four vessels in Section H were displaced sideways by the same pressure wave. One of these vessels was bent by the force of the explosion. The smaller vessels, i.e. the vessels in Sections A, B, C, D, and E were displaced. Twelve smaller cylindrical vessels were found outside the

Figure 9.2 The Pemex site after the explosion. *Source:* Courtesy of Gelling Publishing, Nieuwerkerk aan den IJssel, The Netherlands.

battery limits of the site. Most of these smaller cylindrical vessels exhibited a characteristic failure pattern. A circular front had been torn off. The vessel fragment lacking a front is called an "end tube." The explanation of this phenomenon is as follows. Lines and valves are attached to one front of the vessel. That front is relatively weak and is removed from the vessel by the force exerted by the internal pressure. The vessel is then launched like a rocket. One vessel, having a volume of $45\,m^3$, traveled $1200\,m$.

The specific area in square meter per kilogram product is, for vessels of the same form, inversely proportional to a linear dimension. Thus, the heat input by a fire is, per kilogram vessel content, greater for small vessels than for large vessels. It is possible that this phenomenon explains the greater damage suffered by small equipment than by large equipment at San Juan Ixhuatepec.

Cylindrical vessels will also have suffered from explosions.

It is likely that LPG has left the relatively undamaged cylindrical vessels through relief valves.

Neighboring companies also suffered considerable damage.

Two hundred and seventy houses were severely damaged by the explosions and the fire at San Ixhuatepec. The most extensive damage was done by fire. Vapor/air explosions occurred in some houses.

Remarks

The built-up area and neighboring companies were too close to the site. This accident teaches that the distance between the battery limits and houses and other companies should have been at least $400\,m$. That distance would have saved the majority of the lives that were lost in this accident. It would also have prevented much of the damage suffered by neighboring companies. And $1.5\,km$ would have been an even safer distance as one vessel traveled $1.2\,km$.

A flare at the Pemex site high above the ground with a protection against strong winds would have been better than the ground-level flare present before the accident.

A final remark is that the storage vessels were too close to each other. A bursting vessel could now, probably more than once, involve one or more neighboring vessels in an explosion.

9.4 NIJMEGEN IN THE NETHERLANDS IN 1978

Event

A tank-lorry filled with LPG arrived at a filling station at Nijmegen in The Netherlands at approximately 08.00 h on December 18, 1978 [6].

The total tank volume was $31.7\,m^3$. It was the first discharge of the day for the tank-lorry, so its degree of filling was probably 85%. The tank-lorry thus probably contained $27\,m^3$ of LPG. The driver of the tank-lorry started the transfer of LPG to an open-air storage vessel at 08.23 h. The tank-lorry motor was kept running because the motor drove the transfer pump. Most probably because of an incorrect connection, LPG started to leak and the hot motor ignited a gaseous LPG/air mixture. The fire heated the LPG in the tank-lorry, causing a pressure rise in the tank-lorry hull. At the same time the fire heated the hull material, causing a weakening of that material. Both effects caused bursting of the hull and a warm BLEVE. The driver of the tank-lorry and the operator of the filling station had fled from the filling station by car and remained unharmed. Other personal damage did not occur. The tank-lorry was destroyed by the explosion and the filling station was seriously damaged. Further material damage did not occur. The distance between the filling station and the built-up area was 500 m. The fire-brigade had blocked the area.

The Location and the Tank-lorry

See Figure 9.3. The filling station serves one direction of a four-lane highway. There is a verge between the two lanes of that direction and the two lanes of the other direction. A railway track is at the other side of the highway. The

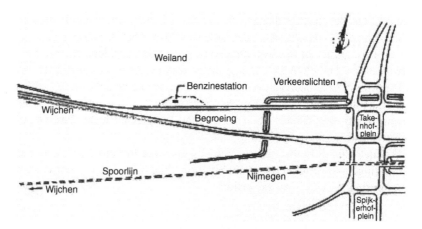

Figure 9.3 Location of the filling station at Nijmegen. A translation of the words in the figure follows: Text of the figure from top to bottom and from the left to the right: Weiland, pasture; Benzinestation, gas station; Verkeerslichten, traffic lights; Begroeing, overgrowth; Spoorlijn, railroad; Wijchen and Nijmegen are, respectively, a village and a town; Takenhofplein and Spijkerhofplein are street names. *Source:* Courtesy of Gelling Publishing, Nieuwerkerk aan den IJssel, The Netherlands.

distance between the tank-lorry and the open-air storage vessel was approximately 10 m. The filling station is surrounded by pastures.

The diameter of the tank of the tank-lorry was 2 m and its length was 10.73 m. The hull's wall thickness was 12.5 mm.

Detailed Description of the Event

The ambient temperature was −4 °C. The saturated vapor pressure of LPG is approximately 4 bara at this temperature. The driver entered the shop of the filling station shortly after 08.20 h. Shortly thereafter, he and the filling station operator noticed a fire under the tank-lorry. They then left the shop to try to extinguish the fire by means of hand-operated fire extinguishers. They concluded on the spot that extinguishing would not be successful because of the size of the fire and the limited amount of extinguishing material. They then fled by car after having asked for the assistance of the fire-brigade by telephone at 08.24 h. The fire-brigade decided to not approach the filling station as there were no humans at the filling station. They blocked the highway and stopped train traffic. The fire-brigade noticed that the relief valve on the storage vessel was relieving and that the relieved material burned. The tank-lorry did not have a relief valve. The tank-lorry burst with a longitudinal crack at 08.45 h and a BLEVE occurred. Remains of the tank-lorry were displaced over 4 m. A front of the tank traveled approximately 50 m and internal partitions of the tank traveled approximately 125 m. A shock wave was not noticed. The fire-brigade went to the site after the explosion and extinguished the fire. The outdoor storage vessel was cooled with water and subsequently emptied by means of hoses into a pasture.

Discussion of the Event

The tank-lorry probably contained 85% of 31.7 m³, i.e 27 m³. The amount of LPG in the tank-lorry was about 14 t. The saturated vapor pressure of liquid propane at 55 °C is 22 bara. The hull of the tank may have burst because of a comparable high pressure and because the fire had weakened the steel of the hull. The specific mass of gaseous LPG exceeds the specific mass of air at the same temperature and pressure by a factor of 1.5–2. So, gaseous LPG is heavier than air and spreads over the ground. The volume of gaseous LPG at ambient temperature and atmospheric pressure exceeds that of liquid LPG by a factor of approximately 250. Because liquid LPG is stored under pressure, liquid LPG flashes when it is exposed to atmospheric pressure. The explosion happened when the tank burst. The volume enlargement by both flashing and combustion caused a sudden pressure increase. See also Sections 9.2 and 9.3.

A Longitudinal Crack

See Figure 9.4. Let D be the tank diameter, L its length, δ its wall thickness, and p the tank pressure. The force that tries to detach a front from the cylindrical part of the tank is $2(\pi/4)D^2p$. The force that keeps the front and the cylindrical part together is $\sigma_1 \pi D \delta$, where σ_1 is the longitudinal material stress. When the tank remains unimpaired, we can equate these two forces. It then follows:

$$\sigma_1 = \frac{2(\pi/4)D^2p}{\pi D \delta} = \frac{Dp}{2\delta}$$

The force that tries to pull the upper horizontal part of the hull from the lower horizontal part is $2DLp$. The force that keeps the upper and the lower parts together is $\sigma_2 2L\delta$. When the tank remains unimpaired, we can equate these two forces. It then follows:

$$\sigma_2 = \frac{2DLp}{2L\delta} = \frac{Dp}{\delta}$$

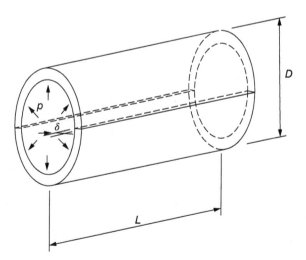

Figure 9.4   A longitudinal crack.

Thus, σ_2, the tangential material stress, exceeds σ_1, the longitudinal material stress, by a factor of two. This fact explains the observation that the tank burst longitudinally.

Remarks

It is safer to transport LPG by rail or by ship than by road transport. If rail transport is planned, it is, whenever possible, better to use rail transport not passing through built-up areas.

The diameter of the fireball has been estimated at 40 m. The center of the fireball has been estimated at 40 m above ground level [6].

The filling station was surrounded by pastures. The nearest built-up area was at a distance of 500 m. Internal parts of the tank traveled 125 m. As we now know it could be advisable to have a distance between LPG filling stations and built-up areas of at least 400 m.

The fire-brigade acted professionally by blocking the highway, stopping train traffic, and starting with extinguishing activities after a BLEVE occurred.

9.5 LOS ALFAQUES IN SPAIN IN 1978

Event

A tank-lorry loaded with 43 m³ of liquid propylene rode on a regional road near the small Spanish town San Carlos de la Rápita on July 11, 1978. It got off the road at 14.35 h due to a bursting tire when it passed the camping Los Alfaques, crossed a ditch, hit a pylon, and broke through the low stone wall of the camping. The accident caused the bursting of the tank and the flashing propylene gave rise to a cold BLEVE. The explosion took the lives of 216 people at the camping. More than 200 people were wounded. There was much material damage.

Detailed Description of the Event

See Figure 9.5. A tank-lorry with a single-hull tank was on its way from Tarragona to Puertollano. The distance between these two towns is 700 km. San Carlos de la Rápita is 90 km south of Tarragona. The tank-lorry had loaded liquid propylene at the harbor of Tarragona. Propylene is the raw material for the manufacture of polypropylene, a plastic. The tank-lorry could have covered the first 250 km on the highway between Tarragona and Sevilla.

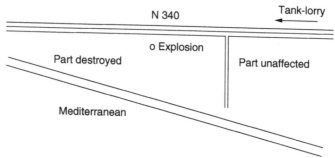

Figure 9.5   Aerial view of the camping Los Alfaques.

However, to avoid having to pay toll, the driver had been instructed to use the regional road. The boiling point of propylene at atmospheric pressure is $-47.70\,°C$. Thus, it can be stored as a liquid at ambient temperature under pressure. So, when the tank of the tank-lorry burst, propylene was a flashing inflammable liquid and gave rise to a BLEVE. The background and the violence of a BLEVE have been discussed in Section 9.2. The explosion may have been ignited by the hot motor of the tank-lorry. Note that the rip up of the tank was caused by a collision and not by a fire. It was a cold BLEVE.

There were 700–800 people present at the camping at the time of the accident. The camping was situated between the regional road and the Mediterranean. The explosion and the following fire hit approximately two thirds of the area, whereas one third remained almost unaffected. The explosion caused fire and subsequently fragmentation of the tank. Fragments were thrown away and one part of the tank crossed the regional road, traveled approximately 300 m, and hit the wall of a restaurant.

The Flashing of Propylene

The flashing of propylene has been discussed in Section 9.2. All liquid propylene in the tank was exposed to the atmosphere when the tank ripped up due to mechanical impact. More than one third of the liquid evaporated almost instantaneously and mixed with air. The evaporating propylene will also have entrained liquid droplets.

Remarks

We see here a worst-case scenario. The accident happened in July and there were many campers on the camping. The driver drove on a regional road. The tank-lorry got off the road at the location of the camping. The tank-lorry had been filled with the maximum volume of 43 m³ liquid propylene. The collision of the tank-lorry caused the bursting of the tank so that the full amount of liquid propylene was immediately available. The ignition source, i.e. the hot motor of the tank-lorry, was directly available.

It would have been better if the tank-lorry had used the highway between Tarragona and Valencia.

The location of the camping along a regional road was not ideal.

It is safer to transport liquid propylene by rail or by ship than by road transport. If rail transport is planned, it is advisable to use rail transport not passing through built-up areas.

There were small containers containing liquid butane at the camping. Many of these containers exploded after the BLEVE occurred.

The tank of the tank-lorry had a single hull. It would have been better if the tank would have had a double hull. However, it is possible that a double hull would not have been able to contain the propylene after the tank-lorry got off the road.

9.6 VIAREGGIO IN ITALY IN 2009

Event

A freight-train consisting of an engine and 14 wagons filled with LPG passed the railway station of Viareggio in Italy on June 29, 2009, shortly before midnight when the shaft of one of the foremost wagons broke. The breakage caused the derailment of the first five or six wagons. The first wagon, directly behind the engine, fell over and hit a small metal pole. That created a hole of approximately 15 cm in the hull [7]. See Figure 9.6. LPG, an inflammable flashing liquid, quickly leaked from the wagon through this hole. Gaseous LPG/air mixture reached a built-up area next to the railway yard. Several vapor/air explosions were ignited in houses, causing the death of 22 people.

Additional Facts

There was, for a relatively short time, a fire around the derailed wagons. However, the derailed wagons did not burst and a BLEVE did not occur. There were, at the location of the derailment, six railway tracks in parallel. The train rode at one side of the railway yard. The gaseous LPG/air mixture crossed the railway yard to reach the built-up area. The specific mass of gaseous LPG exceeds the air specific mass by a factor of 1.5–2. Thus, gaseous LPG spreads over the ground.

Figure 9.6 The damaged LPG wagon at Viareggio. The photograph shows the hole and the pole. *Source:* Courtesy of Gelling Publishing, Nieuwerkerk aan den IJssel, The Netherlands.

Remarks

The accident described was a serious accident. However, the consequences could have been even more serious if a fire had been ignited, reached a wagon, and one or more BLEVEs had occurred. LPG leaked from one wagon only.

It would be wise to avoid rail transport of LPG through town centers.

9.7 A NARROW ESCAPE AT TILBURG IN THE NETHERLANDS IN 2015

Event

A passenger-train, with a velocity of $45 \, \text{km h}^{-1}$, collided head-on with the last wagon of a stationary freight-train at Tilburg in The Netherlands on March 6, 2015 [8]. See Figure 9.7. The last wagon contained 52.5 t of 1,3-butadiene, a flashing inflammable liquid. Personal damage did not occur. The collision caused damage to both the wagon containing 1,3-butadiene and the passenger-train. A small leakage of 1,3-butadiene occurred; however, the escaping material did not ignite. The contents of the wagon could be transferred to a different wagon.

Figure 9.7 Situation after the collision between the wagon and the passenger-train. *Source:* Courtesy of the Dutch Safety Board, The Hague, The Netherlands.

Detailed Description of the Event

The freight-train was on its way from the Chemelot industrial area in the Dutch province Limburg in the southern part of The Netherlands to the shunting-yard Kijfhoek at Zwijndrecht in The Netherlands. Zwijndrecht is approximately 20 km east of Rotterdam. The train consisted of one engine and 35 wagons. Six wagons were loaded and the others were empty. The train had made a stop at the railway-yard Tilburg-Goederen (Tilburg-Goods) between the railway stations Tilburg and Tilburg-Universiteit (Tilburg-University) to enable the replacement of the engine-driver. However, the freight-train was too long for the track on which the train made the stop. The train's length had been increased after the information concerning the train's length had been passed on to the railway company. The freight-train had passed two switches just before it entered the track on which the train came to a halt. On both switches, the freight-train deviated from its original direction. See Figure 9.8. These two switches remained in the position giving access to the track on which the freight-train stood because of the freight-train's length. For the interlocking system, the freight-train had not yet completely passed the two switches.

Rail transport is safeguarded in The Netherlands. The Dutch acronym for the safeguarding system is ATB ("Automatische Treinbeïnvloeding," that is, Automatic Train Control). The objective of the system is to prevent trains from passing stop signs (red signs). It has been mentioned in Section 3.2.3 that the original system has two shortcomings. One shortcoming is that ATB

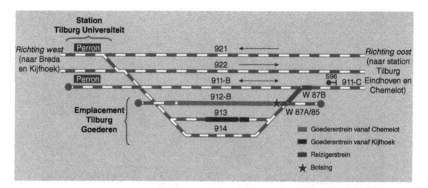

Figure 9.8 The passenger-train passed a stop sign and hit the freight-train on Track 912 B. A translation of the words in the figure follows. Text of the left-hand side of the figure from top to bottom: Station Tilburg Universiteit, railway station Tilburg University; Perron, platform; Richting west (naar Breda en Kijfhoek), direction west (to Breda and Kijfhoek); Emplacement Tilburg Goederen, railway yard Tilburg goods. Text of the right-hand side of the figure from top to bottom: Richting oost (naar station Tilburg, Eindhoven en Chemelot), direction east (to railway station Tilburg, Eindhoven and Chemelot); Goederentrein vanaf Chemelot, freight-train from Chemelot; Goederentrein vanaf Kijfhoek, freight-train from Kijfhoek; Reizigerstrein, passenger-train; Botsing, collision. *Source:* Courtesy of the Dutch Safety Board, The Hague, The Netherlands.

is not active if the train's speed is lower than $40 \mathrm{km\,h^{-1}}$. In the course of the years, the original system has been improved in this aspect. However, the track on which the passenger-train rode prior to the accident had not been improved. The driver of the passenger-train missed a stop sign. ATB did not automatically bring the train to a halt because the train's velocity was $45 \mathrm{km\,h^{-1}}$ and that velocity was not yet high enough to activate ATB. Thus, the passenger-train passed the stop sign and two switches and ran into the freight-train. The driver of the passenger-train activated the brakes just before the collision; however, his train hit the other train while still having a velocity of approximately $45 \mathrm{km\,h^{-1}}$.

Railroad stop signs were passed 169 times in The Netherlands in 2013, so the passing of the stop sign at Tilburg is, in the light of that fact, not exceptional. The passing of a stop sign hardly ever results in an accident in that country because the train's velocity is, in principle, lower than $40 \mathrm{km\,h^{-1}}$.

The passenger-train had been designed around 1964 and was not equipped with crash absorbers. Crash absorbers are parts that, in case of a collision, can absorb energy and thus prevent those parts, which are important for the safety of the passengers or the load, from being damaged. Figures 9.9 and 9.10 depict crash absorbers. The last wagon of the freight-train had been

Figure 9.9   The crash absorbers of the wagon were hardly damaged.
Source: Courtesy of the Dutch Safety Board, The Hague, The Netherlands.

Figure 9.10   This photograph shows damaged crash absorbers (the tubes) that have been damaged by a collision and have absorbed energy. *Source:* Courtesy of the Dutch Safety Board, The Hague, The Netherlands.

equipped with crash absorbers. On the collision, first, the automatic coupling at the front of the passenger-train hit the draw-hook of the wagon. The automatic coupling was pushed in. Second, the front part of the passenger-train hit the crash absorbers of the wagon. Due to height differences, only the upper parts of the crash absorbers were hit. This led the front part of the passenger-train to climb up against the crash absorbers and to hit the dished end of the vessel containing 1,3-butadiene. The crash absorbers were only slightly damaged and thus could only absorb a small amount of energy. See Figure 9.9. Figure 9.10 shows deformed crash absorbers. The latter crash absorbers are not related to the accident at Tilburg on March 6, 2015. The front part of the passenger-train dented the dished end of the vessel containing 1,3-butadiene. The vessel did not rip up; however, the lid of a manhole in the dished end started to leak slightly due to deformation of the dished end. The collision caused a 4-m displacement of the wagon.

It is possible to equip crash absorbers with parts to prevent the other wagon or carriage to climb up against the crash absorbers. See Figure 9.11. It is also possible to equip wagons with vertical shields to protect the dished ends. The intention of the latter provision is to protect the vessel against a penetration by a sharp object. The wagon at Tilburg had not been equipped with these provisions. The installation of a provision to prevent the other wagon or carriage to climb up against the crash absorber or the installation of a vertical shield is obligatory in The Netherlands for wagons transporting a very toxic or pyrophoric liquid or a toxic gas. 1,3-Butadiene does not belong to that category. The Netherlands follows in this respect the rules of the RID [9].

1,3-Butadiene

1,3-Butadiene can be polymerized to polybutadiene. It is an inflammable flashing liquid. Its boiling point at atmospheric pressure is −4.4 °C and the

Figure 9.11   This photograph shows a provision to prevent a wagon or carriage to climb up against a crash absorber. *Source:* Courtesy of the Dutch Safety Board, The Hague, The Netherlands.

saturated vapor pressure at 25 °C is 2.7 bara. At a given temperature and pressure, its specific mass is 1.87 times greater than the air specific mass. So, as a gas, it spreads over the ground and does not rise.

The lower and upper explosion limits (LEL and UEL) at atmospheric pressure and 20 °C are, respectively, 1.4% and 16.3% by volume.

The aforementioned physical properties of 1,3-butadiene illustrate that the accident at Tilburg could well have developed into a cold BLEVE. We see a narrow escape.

Remarks

The Dutch Safety Board recommends that the last wagon of a freight-train should not transport a dangerous good. A further recommendation is to provide crash absorbers of all wagons transporting dangerous goods with parts to prevent a wagon or carriage hitting the wagon containing a dangerous good to climb up against their crash absorbers. Alternatively, to install vertical shields to protect dished ends.

Additional Observations Concerning the Train Accident at Tilburg

ATB and Switches   The damaged freight-train normally rode on a track that is safeguarded by an improved version of ATB (ATB-vv). It means that trains are automatically, irrespective of the train's velocity, brought to a halt if a stop sign is passed. However, the damaged freight-train was not on a track safeguarded by ATB-vv at the time of the collision.

Passing a switch and deviating from the original direction means running a risk. The freight-train had passed two switches on which a deviation from the original direction occurred.

The diversion from the normal track and the passing of two switches with deviations from the original direction occurred to enable the replacement of the engine-driver.

It is not optimum that, because of logistic or economic reasons, additional risks at the transport of dangerous goods are introduced.

The Train's Length   The information concerning the train's length passed on to the railway company was not correct.

The Responsibility of the Companies Transporting Dangerous Goods   The Dutch Safety Board recommends that the companies shipping dangerous goods check the way their goods are transported.

Narrow Escape The collision caused a 4-m displacement of the wagon and of the whole freight-train. This was possible because only 6 of 35 wagons were loaded, and the wagons did not have their brakes activated. The brakes of the engine of the freight-train had been activated. Thus, the freight-train's displacement could absorb part of the collision energy. Such an energy absorption by the freight-train would not have been possible if more wagons had been loaded and if the wagons would have had their brakes activated. More energy to rip up the wagon containing 1,3-butadiene would then have been available.

Final Remarks It has been agreed in The Netherlands to compile trains transporting dangerous goods such that the risk of a warm BLEVE is reduced. It means that a wagon containing an inflammable liquid, such as petrol, will not be next to a wagon containing an inflammable flashing liquid like LPG. The rationale is that if an inflammable liquid takes fire, a wagon next to the damaged wagon will be heated. If that wagon contains an inflammable flashing liquid, a warm BLEVE may develop. See Section 9.2.

9.8 DIEMEN IN THE NETHERLANDS IN 2014

Event

An explosion occurred in the basement of apartment building De Beukenhorst at Diemen in The Netherlands on September 4, 2014 [10]. Work concerning the renovation of an elevator was being carried out that day. A steel transfer line in the foundation of the building was considered a blind one. It was tried to remove the line with a crane. The personnel at the site did not know that a functioning natural gas line was inside the steel line. The attempts to remove the steel line caused, inside the building, damage to the natural gas line, thus allowing natural gas to flow freely into the building. The personnel outside the building did not notice the severe leakage. Explosive natural gas/air mixtures could form in spaces of the building and have subsequently been ignited. The explosion took the lives of two people and wounded 15 people. The material damage was substantial.

Detailed Description of the Event

The activities carried out on September 4, 2014 concerned the removal of the foundations of an elevator-shaft and its entry. See Figure 9.12. The shaft itself had already been removed on that date. The work was carried out by

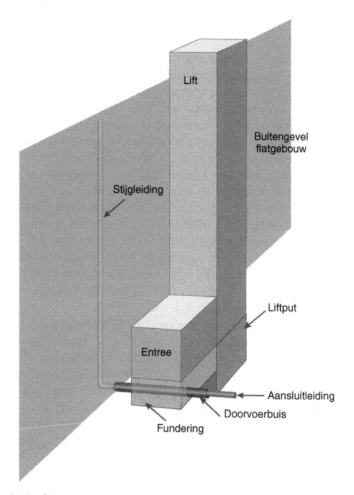

Figure 9.12 Schematic representation of a gas connection at apartment building De Beukenhorst. A translation of the words in the figure follows: Lift, elevator; Buitengevel flatgebouw, outer front apartment building; Stijgleiding, ascending line; Liftput, elevator pit; Entrée, entrance; Aansluitleiding, connecting line; Doorvoerbuis, guiding line; Fundering, foundation. *Source:* Courtesy of the Dutch Safety Board, The Hague, The Netherlands.

three men, including a crane-driver, working for a subcontractor. The crane-driver, while removing rubble with the crane, noticed a steel line below ground level at approximately 15.00 h. The line ran perpendicular to the apartment building. The three men thereupon stopped working. The crane-driver reported his finding to the general foreman and asked for instructions. The general foreman and the supervisor of the housing association checked the situation and then consulted drawings in the site hut. The drawings did not show the line. The general foreman then drew the conclusion that the line was a blind one and ordered the crane-driver to remove the line. The crane-driver detached the line from the foundation at the street side with the crane at approximately 15.15 h. The three men then noticed a yellow line inside the steel line, which they identified as a natural gas line. They reported their finding to the general foreman who again checked the situation. Those present at the site smelled natural gas, however, only faintly. They drew the conclusion that a small amount of gas was released by the steel line. The general foreman called the gas distributing company at 15.27 h to report the incident. The mention was classified as "urgent," not as "very urgent." The employee of the gas distributing company told the general foreman that a fitter would come to check the situation.

Then, the work was stopped and the crew waited for the fitter. The general foreman and the three men working for the subcontractor remained at the site. The supervisor of the housing association and one more employee of the same organization were also present. The site was not evacuated. Neither the fire-brigade nor the police was informed.

Several residents smelled gas leakage: two of them reported this to the employees at the site and one resident called the housing association. The employee of the housing association present at the site entered the basement of the building. She told those present that she smelled gas strongly and heard the sound of escaping gas. The crane-driver stood behind her in the doorway. A natural gas/air explosion occurred at 15.41 h and killed both the employee and the crane-driver. Fifteen people were wounded. The general foreman, the supervisor of the housing association, and several residents were among the wounded. The explosion caused considerable damage to the building. Part of the building took fire.

Cause of the Explosion

The yellow natural gas line noticed by the personnel of the subcontractor transported natural gas from a main line to a number of apartments. The detachment of the steel line caused damage to the yellow line inside the basement of the building. A coupling broke loose and the natural gas could flow freely into the building. The lower explosion limit (LEL) of natural gas in air at atmospheric pressure and ambient temperature is 5.0% by volume. The upper explosion limit (UEL) is 16.0% by volume. Natural gas/air mixtures had formed in the basement of the building. Part of these mixtures contained between 5.0% and 16.0% by volume of natural gas in air. Such mixtures had been ignited.

Prologue of the Accident

See the timeline of the accident in Figure 9.13. The housing association possessing De Beukenhorst, built in 1974, decided to renovate the elevators of the building in 2012. It was necessary to upgrade the elevators. An important aspect was that new elevators should be larger than the existing ones. A contractor signed the order for the renovation on February 12, 2014. It would be necessary to dig into the ground because the new foundation would be larger than the existing foundation. It is, in such a case, obligatory in The Netherlands to collect information concerning the presence of cables and lines in the ground. The contractor informed a central organization in The Netherlands and asked for information on March 14, 2014. The information was received; however, the presence of a natural gas line between the main natural gas line and the apartment building at the location of the elevator-shaft in question was not communicated. Nevertheless, that gas line was present in the spring of 2014. The contractor was instructed by the natural gas distributing company to coordinate his work with that company. There was a communication between the contractor and the natural gas distributing company on March 19, 2014. As stated, a mention of the work had been made on March 14, 2014. The information received thereupon had a validity of 20 days. The work had to be started within that period; otherwise a new mention would have to be made. The reason for this limited validity is that the situation might have

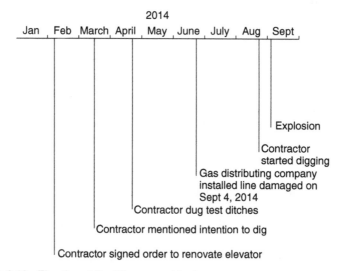

Figure 9.13   Timeline of the Diemen accident.

changed after 20 days. A guideline "Preventing damage to cables and lines due to digging" is valid in The Netherlands. In line with this directive, the contractor dug test ditches to check exactly the position of cables and lines on April 25, 2014. The contractor then concluded that a main natural gas line had to be relocated to allow the building of the new, enlarged elevator-shaft. The main natural gas line had been made of gray cast iron. The gas distributing company planned to replace that line in the fall of 2014. The reason for this replacement was that gray cast iron is vulnerable to vibrations and external loads. The plan was changed because of the renovation of the elevator-shafts. First, the gas distributing company had the natural gas lines between the main line and the apartments replaced in the months April through June 2014. The line damaged on September 4, 2014 was replaced on June 23, 2014. Next, the company had the main line replaced between July 7, 2014 and August 27, 2014. The company informed the contractor that the renovation of the elevator-shafts could be started after August 24, 2014. The contractor started the work shortly after that date. The contractor did not renew the digging mention. The earlier digging mention had lost its validity as more than 20 days had elapsed. Had he done so, information regarding the line damaged on September 4, 2014 would have been received. The reason for not renewing the mention is that the contractor thought he already possessed all relevant information because of his communications with the gas distributing company.

Observations

Damage to natural gas lines due to digging activities is a frequently occurring phenomenon in The Netherlands. For example, 5207 of these cases have been registered in that country in 2013. Approximately 75% of these cases concern damage to lines between main lines and houses, apartment buildings, offices, and so on. Damage to these lines may result in an accumulation of natural gas inside buildings and that is potentially dangerous.

An aspect was that digging in the ground was not a core activity for the contractor of the renovation of the elevator shaft.

A principal of building activities is, according to Dutch law, obliged to see to it that digging activities on behalf of his premises are carried out carefully. The housing association was the principal in this case. The Dutch Safety Board concludes that the principal could have done more to ensure that the contractor would collect all relevant information concerning underground natural gas lines.

The system to show all cables and lines to contractors making a digging mention in The Netherlands can be improved. The natural gas line in the foundation of the elevator-shaft was not shown in the spring of 2014. The Dutch Safety Board recommends to improve that system.

The prologue of the accident shows that the contractor initially took steps and measures to deal carefully with natural gas lines. The renovation of the elevator was delayed by months because a main natural gas line had to be relocated.

The guideline "Preventing damage to cables and lines due to digging" prescribes to stop the work when an unknown line is met. The work should not be resumed until the purpose of the unknown line has been established.

The same guideline orders the evacuation of the site, removal of ignition sources, and calling the fire-brigade or the police when a natural gas line has been damaged. The reason that these actions were not taken is that, outside the building, it did not appear that the situation called for these measures. The leakage was inside the building and was not noticed.

The operator of the gas distributing company receiving the call on the damage classified the situation as "urgent" and not as "very urgent." The Dutch Safety Board recommends to improve the procedure on accepting damage reports.

REFERENCES

[1] Pietersen, C.M. (2009). *After 25 Years – The Two Largest Industrial Disasters – LPG Disaster Mexico City – Bhopal Tragedy*, 48. Nieuwerkerk aan den IJssel, The Netherlands: Gelling Publishing (in Dutch).

[2] Pietersen, C.M. (2009). *After 25 Years – The Two Largest Industrial Disasters – LPG Disaster Mexico City – Bhopal Tragedy*, 17–22, 31–62, 117–123. Nieuwerkerk aan den IJssel, The Netherlands: Gelling Publishing (in Dutch).

[3] Pietersen, C.M. (1988). Analysis of the LPG-disaster in Mexico City. *Journal of Hazardous Materials* 20: 85–105.

[4] Arturson, G. (1987). The tragedy of San Juanico – the most severe LPG disaster in history. *Burns* 13: 87–102.

[5] Pietersen, C.M. (2009). *After 25 Years – The Two Largest Industrial Disasters – LPG Disaster Mexico City – Bhopal Tragedy*, 38. Nieuwerkerk aan den IJssel, The Netherlands: Gelling Publishing (in Dutch).

[6] Steunenberg, C.F., Hoftijzer, G.W., and van der Schaaf, J.B.R. (1981). Investigation concerning an accident at Nijmegen at which a tank-lorry was involved. *Polytechnisch tijdschrift/procestechniek* 36: 175–182. (in Dutch).

[7] Pietersen, C.M. (2009). *After 25 Years – The Two Largest Industrial Disasters – LPG Disaster Mexico City – Bhopal Tragedy*, 169. Nieuwerkerk aan den IJssel, The Netherlands: Gelling Publishing (in Dutch).

[8] Dutch Safety Board (2016). *Risk Control at Rail Transport of Dangerous Goods*, 1–102. The Hague, The Netherlands: Dutch Safety Board (in Dutch).

[9] Convention relative aux transports internationaux ferroviaires (COTIF) (2011). *Appendice C – Règlement concernant le transport international ferroviaire des marchandises dangereuses (RID)*, 1-1–7-19. France: COTIF (in French).

[10] Dutch Safety Board (2015). *Dangers of Gas Lines on Digging*, 1–70. The Hague, The Netherlands: Dutch Safety Board (in Dutch).

10

NUCLEAR POWER STATIONS

10.1 INTRODUCTION

10.1.1 General

There are four aspects concerning the public acceptance of nuclear energy: severe accident risk, proliferation, vulnerability to sabotage, and nuclear waste disposal. Severe accident risk in nuclear power stations is treated in this chapter. The reason for this choice is that, until now, only reactor accidents have had serious consequences.

In 2013, approximately 16% of the electricity production worldwide and one third in the European Union was obtained from nuclear fission [1]. In 2010, 62% of the nuclear reactors in power stations worldwide were pressurized water reactors (PWRs) and 19% were boiling water reactors (BWRs) [2]. Both reactor types belong to the category of light water reactors (LWRs). These reactor types will be described in some detail in Section 10.2 as serious accidents occurred with these reactor types. Still, they are often considered to be options for the future. Nuclear power stations were introduced in the 1950s. Until now, three major

Safety in Design, First Edition. C.M. van 't Land.
© 2018 John Wiley & Sons, Inc. Published 2018 by John Wiley & Sons, Inc.

accidents occurred in large nuclear power stations. First, in 1979, there was the TMI-2 accident at Harrisburg, Pennsylvania, USA. It concerned a PWR. The consequences were modest. The accident is discussed in Section 10.3. Second, in 1986, an accident occurred at Chernobyl in Russia. The consequences of the accident at Chernobyl were very serious. The reactor type was different from either a PWR or a BWR, and the accident is not discussed in this book. Generally, that reactor type is not considered an option for the future as its safety characteristics are not optimum. Third, in 2011, there was an accident at Fukushima in Japan. It concerned three BWRs. The consequences were serious. What happened with the BWR of Unit 1 of the nuclear power station Fukushima Daiichi is described in Section 10.4.

Both at Harrisburg and Fukushima, primary process protection was by means of active and procedural safety measures. Two reactor types having passive safety as primary process protection are discussed in Section 10.5. It concerns the pebble bed reactor (PBR) and the prismatic block reactor. The reason for discussing these two reactor types and not, e.g. the molten salt reactor, is that large power stations have functioned with these two reactor types, albeit with mixed results. THTR-300 in Germany, having a capacity of 300 MWe(lectrical), operated for 4 years with a PBR. Fort St. Vrain Unit No. 1 in the United States, having a capacity of 330 MWe, functioned for 10 years with a prismatic block reactor.

Finally, these two reactor types are compared in Section 10.6. It is recommended to further explore the possibilities of the prismatic block reactor.

10.1.2 Physics

Matter consists of atoms. A specific type of atom is characteristic for a chemical element. In many materials, atoms are combined to form molecules. The properties of a molecule are characteristic for that material. It has been for a long time considered that atoms cannot be split up and thus are the smallest particles of matter.

Mendeleev arranged the chemical elements in the Periodic Table of Elements. At present, 118 elements are incorporated into this Periodic System. Hydrogen is found in the upper-left corner of that scheme. The hydrogen atom is the smallest atom. The famous Danish scientist Niels Bohr proposed a model for the hydrogen atom, which is basically still used today. It consists of a positively charged nucleus, called a proton, and a small, negatively charged, electron. The electron rotates around the nucleus, and the force with which these two particles attract each other is counterbalanced by the centripetal force (see Figure 10.1).

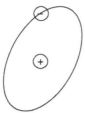

Figure 10.1   The hydrogen atom.

The atoms of deuterium and tritium are related to the hydrogen atom. The nucleus of a deuterium atom contains, in addition to a proton, an uncharged, neutral particle having the same mass as a proton. It is called a neutron. The nucleus of a tritium atom contains, in addition to a proton, two neutrons. All three atoms have one electron rotating around its nucleus. Hydrogen, deuterium, and tritium are called isotopes. Isotopes have the same chemical properties because their electronic configurations are equal. However, their atomic masses are different, and that fact provides possibilities to separate them physically. Other elements also have isotopes. In this context, it is useful to distinguish between light water and heavy water. Normal water is light water. The hydrogen atoms of light water have a nucleus that consists of one proton. One electron rotates around the nucleus. The deuterium atoms of heavy water have a nucleus that consists of one proton and one neutron. One electron rotates around the latter nucleus. In the molecules of both light water and heavy water, two hydrogen or two deuterium atoms are combined with one oxygen atom.

Relatively light atomic nuclei contain approximately equal numbers of protons and neutrons. For example, the nucleus of a nitrogen atom contains seven protons and seven neutrons. It is customary to provide the chemical symbol of an element with two numbers: the mass number and the atomic number. The mass number equals the sum of the number of protons and neutrons in an atomic nucleus, whereas the atomic number equals the number of protons, e.g. $^{14}_{7}N$ for nitrogen.

The number of electrons equals the number of protons in a neutral atom. On moving from relatively light atoms to heavier atoms, the number of neutrons in the nuclei of those atoms becomes larger than the number of protons. If, in an atomic nucleus, the number of neutrons exceeds the number of protons significantly, the distance between the protons increases and the nucleus loses stability. The element bismuth (Bi) has the heaviest atomic nucleus that is still stable. The number of protons in its nucleus is 83, whereas the number of neutrons in its nucleus is 126. Thus, the mass number is the sum of these two numbers, i.e. 209, and the atomic number is 83.

The majority of nuclear power stations use uranium (U) as fuel. In nuclear reactors, uranium is present in the oxide, UO_2, or in the carbide. Uranium occurs in an ore having the chemical formula U_3O_8, meaning a combination of three uranium atoms and eight oxygen atoms. The atomic nucleus of U contains 92 protons and its atomic number is 92. Naturally occurring uranium contains three isotopes of U and their atomic masses are 234, 235, and 238. It is customary to indicate them as U-234, U-235, and U-238. Naturally occurring uranium contains a very small and negligible amount of U-234. It consists of 99.3% by weight of U-238 and 0.7% by weight of U-235. The nuclei of both isotopes are unstable and exhibit radioactivity; they decay and emit radiation. Decay means that other nuclei are formed out of these uranium nuclei. The sum of the masses of the nuclei formed is less than the mass of the original nucleus. The difference is emitted as radiation energy according to the famous formula $E = mc^2$. E is the energy in J, m is the mass difference in kg, and c is the velocity of light in $m\,s^{-1}$. The radiation energy can be converted into heat and that is what occurs in nuclear power stations.

U-238 is much stabler than U-235. Technically, it is possible to split U-235 to produce energy. It is difficult to split U-238. In order to be effective as fuel in nuclear power stations, the percentage by weight U-235 of the sum of the masses of U-235 and U-238 must, for the majority of nuclear reactors, increase to at least 3–5. That increase is called enrichment and can be achieved by physical processes. The process used widely today is enrichment by means of ultracentrifuges. The internationally accepted maximum degree of enrichment in use today is 20% by weight. The reason is the necessity to avoid proliferation.

Thorium can also be used as a nuclear fuel. Thorium is a naturally occurring element having an atomic number 90 and a mass number 232, hence often indicated as Th-232. It can be used as nuclear fuel in combination with U-235. Worldwide, there is much more thorium available than uranium. In nuclear reactors, thorium is present as the oxide, ThO_2, or as the carbide.

10.2 PRESSURIZED WATER REACTORS (PWRs) AND BOILING WATER REACTORS (BWRs)

10.2.1 Introduction

The common characteristics of these two reactors are dealt with in this section. They both belong to the category of LWRs. Light water is normal water, and the difference between light water and heavy water has been mentioned in Section 10.1.2. Both a PWR and a BWR is a vessel having a thick steel wall and containing water, either as liquid (PWR) or as liquid and vapor (BWR), and fuel in metal tubes; see Figure 10.2 that depicts a BWR. The most used fuel is uranium dioxide (UO_2). The physical form of the dioxide is a sintered pellet having a diameter of approximately 1 cm and a height of approximately 1 cm. The pellets are stacked one on top of the other in fuel tubes. Typically, the uranium in the oxide consists of 5% by weight of U-235 and 95% by weight of U-238. The metal tubes are close to each other to reach the critical mass and to enable nuclear fission. U-235 decays, neutrons are emitted, and heat is produced. The neutrons emitted cause further fission and heat production.

Moderation of the neutrons is an important function of water in the reactor. Moderation is the deceleration of the emitted neutrons from approximately $14\,000\,\mathrm{km\,s^{-1}}$ to approximately $2.2\,\mathrm{km\,s^{-1}}$ [3]. Only these relatively slowly moving neutrons can accomplish further fission of U-235. Another key function of water is the absorption of heat to produce mechanical and subsequently electrical energy. Steam is raised in the reactor shown in Figure 10.2 and passed on to a turbine. The turbine rotates and drives an electricity generator. The condensed steam is returned to the nuclear reactor.

On moving the control rods in a vertical direction between the fuel tubes, the fission process can be controlled. The metal or alloy of the

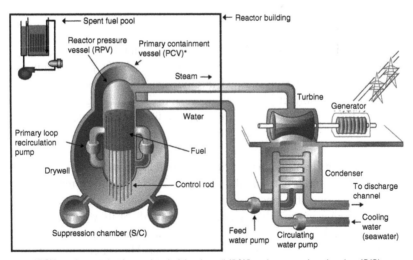

*PCV: equipment that is consisted of the drywell (D/W) and suppression chamber (S/C)

Figure 10.2 A nuclear power station equipped with a BWR. *Source:* Courtesy of TEPCO, Tokyo, Japan.

control rods absorbs neutrons. The nuclear fission is halted by moving the control rods fully between the fuel tubes. However, the decaying radioactive materials left from the fission continue to heat the reactor's coolant water. These decaying radioactive materials are called actinides. Immediately after the fission is halted, the heat in MW raised in the reactor is approximately 7% of the heat in MW raised before the fission was halted. The percentages are approximately 2, 0.5, and 0.1 after, respectively, an hour, a day, and a year [4].

When a nuclear reactor produces at, e.g. 100% capacity, 93% of the heat development is due to the fission of U-235. The remaining 7% of the heat development is due to the decay of actinides. The heat development due to the fission of U-235 can be stopped completely by the introduction of control rods. However, the heat development due to the decay of actinides cannot be stopped by the introduction of the control rods. Thus, the heat development due to the decay of actinides is not something that occurs when the control rods are introduced. This heat development is an integral part of the heat produced by a nuclear reactor when the reactor produces normally.

The reactor is placed in a building with concrete walls having a thickness of one to several meters to contain radiation. The reactor is called critical when it reaches a stationary state and the nominal heat flow is produced.

10.2.2   PWR

Figure 10.3 is a diagram of the nuclear power plant TMI-2 at Harrisburg. An accident in this plant is discussed in Section 10.3. The power plant is no longer active. The water pumped through the reactor does not boil and exchanges heat indirectly in steam generators. The steam generated flows to turbines and brings about their rotation. Radioactive material thus cannot reach the turbines when a fuel tube is damaged. The turbines effect the rotation of electricity generators.

Typically, the pressure in the primary circuit is 155 bar, whereas the temperature of the water entering the reactor is 290 °C and the temperature of the water leaving the reactor is 325 °C. A pressurizer serves to prevent boiling of water in the primary circuit. Two thirds of its volume are filled with water and one third is vapor space. Water having a temperature of 345 °C has a saturated vapor pressure of 155 bar. The water temperature of 345 °C in the pressurizer is maintained by either heating water electrically or by adding cold water. So there is a margin of 20 K to prevent boiling of the water flowing to a steam generator. Typically, saturated steam having a

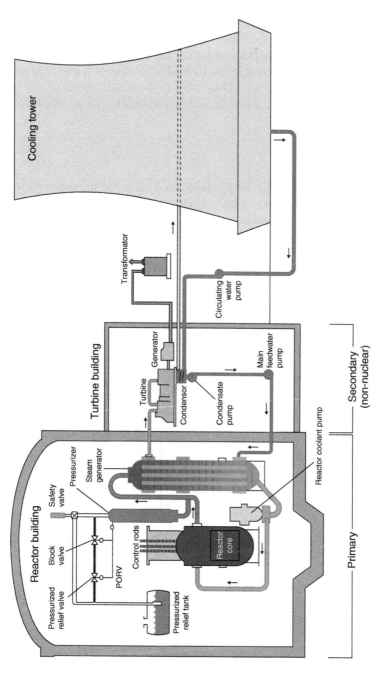

Figure 10.3 The nuclear power station TMI-2 at Harrisburg in the United States. *Source:* Courtesy of U.S. Nuclear Regulatory Commission, Washington, DC, USA

pressure of 60 bar and a temperature of 275 °C is raised in the steam generators. It is passed on to the turbines, condensed, and recycled to the steam generators. The reactor, the steam generators, the pressurizer, and the circulation pumps form the primary circuit and are inside the containment. A typical PWR having a capacity of 1000 MWe(lectrical) is equipped with two or four steam generators. MWe refers to the actual output of electric power, the heat raised in a PWR is, as MWth(ermal), a factor of approximately 2.5 larger.

10.2.3 BWR

See Figure 10.4. The steam for the turbines is raised in the reactor itself. Thus, the design of a BWR is simpler than that of a PWR. Water circulation through the reactor is accomplished by, e.g. two pumps. The water boils in the reactor. There are steam bubbles in the upper part of the reactor.

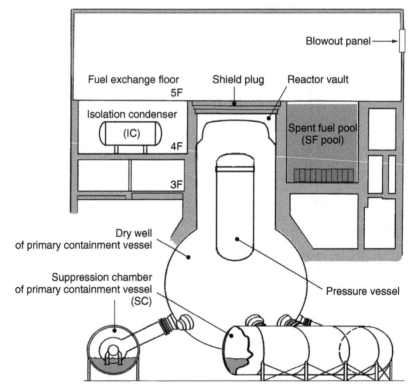

Figure 10.4 Fukushima Unit 1 – a nuclear power station with a BWR.
Source: Courtesy of Springer Japan, Tokyo, Japan.

Moderation is hence more successful in the lower part of the reactor than in the upper part. That is an important reason why the control rods pass through the bottom of the reactor (see also Figure 10.2). Steam leaving the reactor entrains droplets and devices in the reactor's vapor space separate droplets from steam. Dry steam passes to the turbines and separated water is recycled to the reactor. Typically, the saturated steam has a pressure of 75 bar and a temperature of 290 °C. After having passed the turbines, the steam is condensed and water is recycled to the reactor. Radioactivity can reach the turbines when a fuel tube is damaged.

10.3 THREE MILE ISLAND (TMI)

Event

An accident concerning the reactor of a nuclear power station occurred at Harrisburg on March 28, 1979. The nuclear power station was located on Three Mile Island, and the accident is often indicated by the acronym TMI. The site at Three Mile Island consisted of two nuclear power stations and the accident occurred in TMI-2. The other nuclear power station is indicated as TMI-1. Due to a loss of cooling capacity in the reactor of TMI-2, the temperature in the reactor increased and caused the melting of a number of fuel rods. A chemical reaction between zirconium in the fuel rods and water generated hydrogen. Hydrogen entered the containment and caused an explosion. The containment was not breached. The major release of radioactivity on the morning of March 30, 1979, was caused by the controlled, planned venting of the make-up tank into the vent header. The header was known to have a leak [5]! However, the release did neither harm the public nor the employees of the power stations [6]. TMI-2 was severely damaged.

TMI-2

The capacity of the two nuclear power stations together was 1700 MWe. The reactor of TMI-2 was a PWR (see Figure 10.3), and the pressure in the primary circuit was 2150 psi (146.3 bar). Water having a temperature of 340 °C has a saturated vapor pressure of 2150 psi. The water in the pressurizer had that temperature. Probably, the water leaving and entering the reactor had temperatures of, respectively, 320 and 290 °C. The primary circuit consisting of the reactor, the pressurizer, two steam generators, and four reactor coolant pumps can be distinguished. UO_2-pellets were present in thin tubes made of Zircaloy-4 and having a length of 12 ft (3.66 m). The walls of the thin tubes are called "cladding". Zircaloy-4 is an alloy containing the metal zirconium. The primary circuit was located in the containment indicated as the reactor building. The reactor building had walls of reinforced concrete having a thickness of 4 ft (1.22 m). The control rods passed through the top of the reactor. The secondary circuit was mainly outside the reactor building. It consisted of a turbine, a condenser, pumps, and a cooling tower. Heat was exchanged indirectly in the steam generators.

Additional Remarks Concerning TMI-2

At roughly 2200 °F (1204 °C), a reaction between the Zircaloy cladding and water could begin to damage the fuel rods and also generate hydrogen. Damage to the cladding releases some radioactive materials trapped inside the fuel rods into the core's cooling water.

At about 5200 °F (2871 °C), fuel starts to melt. Such melting could release far more radioactive materials than the damage done to the fuel rods at 2200 °F.

Description of the Accident

Reference 7 contains a description of the accident and Table 10.1 a timeline. TMI-2 ran at 97% power prior to the accident. It started at 04.00 h by the tripping of the first pump of a series of feedwater system pumps supplying water to the steam generators. Within 1–2 s, all feedwater pumps tripped. These pumps were part of the secondary circuit. A "trip" means a piece of machinery stops operating. The cause of the tripping was probably the leaking of water into instrument lines. Such leaking had occurred at least twice earlier at TMI-2. When the pumps stopped, the flow of water to the steam generators stopped. The plant's safety system automatically shut down the steam turbine and the electricity generator it powered.

When the pumps that normally supply water to the steam generators shut down, three emergency feedwater pumps automatically started. However, they could not deliver water to the steam generators as valves in the two emergency feedwater lines were closed. The stopping of the feedwater flow to the steam generators caused an increase in the temperature of the water in the pressurizer. The pressure in the vapor space of the pressurizer rose to 2255 psi (153.4 bar). Water having a temperature of 340 °C has a saturated vapor pressure of 146.3 bar. Water having a temperature of 343.5 °C has a saturated vapor pressure of 153.4 bar. Thus, a relatively

Table 10.1 Timeline of the accident at Three Mile Island.

The accident started at 04.00 h on March 28, 1979	
Time into the accident	**Action**
0 s	Trip of a feedwater pump
1–2 s	Trip of the other feedwater pumps
1–2 s	Turbine and generator shut down
Several seconds	Emergency pumps start – no water
Several seconds	Relief valve opens
8 s	Control rods into the reactor
13 s	Relief valve remains stuck open
14 s	Operator errs about emergency pumps
2 min	HPI pumps start
4.5 min	Operator reduces flow from HPI pumps
1 h and several minutes	Coolant pumps start to vibrate severely
1 h and 14 min	Operator shut down two coolant pumps
1 h and 31 min	Operator shut down the other two coolant pumps
2 h and several minutes	Personnel notes open relief valve
2 h and 22 min	Valve next to relief valve closed
3 h and 22 min	High-pressure water injection resumed
9 h and 50 min	Explosion in reactor building
About 18 h	Reactor stable
About 50 h	Release of radioactive gases

small temperature increase leads to a substantial pressure rise. A relief valve atop the pressurizer (PORV) opened several seconds into the accident and steam and water began flowing out of the reactor coolant system to a relief tank. However, pressure continued to rise and 8 s after the tripping of the first pump the control rods automatically dropped down into the reactor core to stop the nuclear fission. The pressure continued to rise because, when the relief valve opened, the reactor was still running at 97% capacity.

The remaining heat development of the reactor after the control rods had dropped in was still 6% of the heat development prior to the start of the accident. With the reactor shut down and the relief valve open, pressure in the reactor coolant system fell. The pressure fell because the reactor was now, as the control rods had dropped into the reactor, at 6% capacity. Up to this point, the reactor system was responding normally to a turbine trip. The relief valve should have closed 13 s into the accident, when pressure dropped to 2205 psi (150 bar). It did not. A light on the control room panel indicated that the electric power that opened the relief valve had gone off, leading the operators to assume the valve had shut. But the relief valve was stuck open and would remain open for 2 h and 22 min, draining needed cooling water. In the first 100 min of the accident, some 32 000 gallons (121 m³), over one third of the entire capacity of the reactor's primary cooling system, would thus escape. Fourteen seconds into the accident, an operator in TMI-2's control room noted that the emergency feedwater pumps were running. He did not notice two lights that could have told him that a valve was closed in each of the two emergency feedwater lines. Water could not reach the steam generators. One light was covered by a yellow maintenance tag. No one knows why the second light was overlooked.

Two minutes into the accident, pressure in the primary circuit dropped sharply. Automatically, two high-pressure injection (HPI) pumps began pouring about 1000 gallons per minute (227.1 m³ h⁻¹) into the primary circuit. At the same time, because of the low pressure in the primary circuit, the water in this circuit started to boil. The level in the pressurizer rose and the operators took it that there was enough water in the primary circuit. However, the level was high because of the presence of steam bubbles in the water. About 4.5 min into the accident, an operator shut one HPI pump down and reduced the flow of the second HPI pump to less than 100 gallons per minute (22.7 m³ h⁻¹).

Slightly more than an hour into the accident, TMI's four reactor coolant pumps began to vibrate severely. The vibrations were caused by the pumping of both water and steam. The HPI water supply to the primary circuit was too small relative to the water removal as steam via the relief valve due to the heat development of the fuel. Even so, after 1 h and 14 min, it was decided to close down two of the four reactor coolant pumps. After 1 h and 31 min, the two remaining pumps were shut off.

As from the latter point in time, the fuel rods were no longer submerged in water and the temperature of the fuel rods increased. A reaction between water and zirconium occurred and hydrogen was generated. Hydrogen escaped through the open relief valve into the reactor building. It mixed with air containing oxygen and a vapor explosion occurred 9 h and 50 min after the start of the accident [8]. The pressure increase due to this vapor explosion was 28 psi (1.9 bar). The containment was not breached by the explosion.

Slightly more than 2 h into the accident, it was established that the relief valve was open and after 2 h and 22 min an adjacent valve was closed to stop the leaking of water and steam into the reactor building. Still 1 h later, high-pressure water injection was resumed. By the evening of March 28 (the day of the accident), the core appeared to be adequately cooled and the reactor appeared to be stable.

The major release of radioactivity on the morning of March 30, 1979, was mentioned in the beginning of this section. Remember that the accident started at 04.00 h on March 28, 1979.

Remarks

The heart of the matter of the accident at Three Mile Island is that it has not been possible to transfer the remaining heat of the nuclear reactor in a safe way.

The protection of the nuclear reactor at Three Mile Island relied on passive, active, and procedural safety measures (see Chapter 2). A passive safety measure is, within very wide limits, not endangered by human errors or equipment failure. The building provided passive protection by preventing the emission of radioactivity. Active process protection starts working upon a signal. Procedural safety measures concern action to be taken by humans.

The first active safety measure to be discussed was the automatic shutdown of the turbine and the generator it powered. This measure functioned well.

The second active safety measure to be mentioned was the automatic starting of three emergency feedwater pumps. This action was started by the tripping of the feedwater pumps. The back-up pumps could not deliver water to the steam generators because valves were closed in their delivery lines. This protection method failed. Fourteen seconds into the accident, an operator noted that the three emergency feedwater pumps were running. However, it was not noted that valves in the delivery lines were closed.

The third active protection method was the opening of a relief valve atop the pressurizer. The relief valve was opened by a high-pressure signal from the pressurizer when less than 8 s had elapsed after the start of the accident. This measure functioned well. However, the relief valve should have closed when 13 s into the accident had elapsed because the pressure in the pressurizer had come down. The relief valve did not close. The third safeguarding method failed as well.

The fourth active safeguarding method to be mentioned is the automatic dropping of the control rods into the core to halt nuclear fission. This action was started because the pressure in the reactor continued to rise after the relief valve had opened. This protection method functioned satisfactorily. Indeed, the control rods fell into the core 8 s into the accident. However, this action reduced the heat flow from the reactor to 6% of the nominal value and not to 0%.

The fifth active protection method was the automatic starting of two HPI pumps 2 min into the accident. The signal starting this action is low pressure in the primary circuit. This measure functioned well. However, the action of the HPI pumps was mitigated by human action. The final result was the failing of this protection method as well.

None of the primary process protection measures was a passive safeguarding method.

Damage

Personal damage did not occur. The small radioactive releases had no detectable health effects on plant workers or the public [6]. The nuclear power station TMI-2 was severely damaged.

10.4 FUKUSHIMA UNIT 1

Event

A nuclear accident occurred at Fukushima in Japan on March 11, 2011. It concerned the nuclear power station Fukushima Daiichi. The word "Daiichi" means "Two" in Japanese. There were two nuclear power stations at Fukushima and the accident occurred at the second station. Both stations were operated by TEPCO (Tokyo Electric Power Corporation). The Fukushima Daiichi station consisted of six nuclear power plants, each indicated by the word "Unit" followed by a number. At the time of the accident, Units 4–6 were not operating due to periodical inspection.

Following a major earthquake, a 15-m tsunami disabled the power supply and cooling of the reactors of Fukushima Daiichi Units 1–3. The cores of these three units largely melted in the first 3 days. Unit 1 will be in focus mainly in the next sections.

Unit 1

Unit 1 was completed and started to operate in March 1971 (see Figure 10.4). It was the third BWR built in Japan, and most of the design and manufacture of the key components were done by General Electric (GE). Its electrical output was approximately 460 MWe. The operational conditions resembled the conditions outlined in Section 10.2 for a BWR. The steam passed on to the turbines was raised directly in the reactor. The reactor was equipped with two circulation pumps, indicated as Pumps A and B.

The vessel shaped like a lightbulb was the primary containment vessel (PCV). The containment of Unit 1 is indicated by the manufacturer as MARK I. The suppression chamber (SC) of the PCV contains cold water and serves to condense steam that would have escaped from the reactor.

Additional Notes Concerning Fukushima Daiichi Unit 1

The melting point of the core fuel, i.e. UO_2, is 2880 °C [9]. The melting point of a mixture of UO_2, ZrO_2, and Zircaloy is 2000–2200 °C. The melting point of stainless steel is approximately 1500 °C.

Detailed Description of the Event

Reference 10 contains a description of the accident and Table 10.2 a timeline. An earthquake hit the power station Fukushima Daiichi at 14.46 h on March 11. The output of the power station was approximately 460 MWe at this point in time. The earthquake caused a loss of electric power and diesel generators started automatically to supply electric power. Also automatically, the main steam isolation valve was closed and the control rods passed into the reactor to stop nuclear fission. The reactor's heat output was thereby reduced to 7% of the heat output at the time the earthquake

Table 10.2 Timeline of the events in Unit 1 of Fukushima Daiichi.

The accident started at 14.46 h on March 11, 2011	
Time into the accident	Action
0 s	Earthquake hits the power station
Several seconds	Diesel generators start
	Main steam isolation valve closes
	Control rods into the reactor
6 min	ICs start
17 min	Operators stop IC-cooling
About 27 min	Manual on/off-control of one IC started
About 50 min	Operators close valve MO-3A
53 min	Tsunami, IC cooling lost
8 h 14 min	Reactor leakage established
About 9 h	Reactor depleted of water
13 h and 14 min	Seawater pumping into the reactor starts
23 h and 44 min	Pressure of PCV relieved
24 h and 50 min	Explosion in the fuel exchange floor
2 wk	Reactor stable

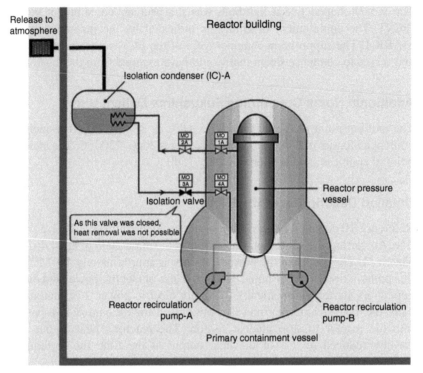

Figure 10.5 Fukushima Unit 1 – Isolation Condenser A. *Source:* Courtesy of Springer Japan, Tokyo, Japan.

occurred. As the main steam isolation valve was closed, the water in the reactor was heated and the pressure rose. Six minutes into the accident, two isolation condensers (ICs) were activated by a signal caused by a high pressure in the reactor (se Figure 10.5). Unit 1 then switched to cooling by ICs A and B. Steam raised by the remaining heat flowed to coils in the ICs. It condensed in the coils and water flowed back into the reactor. ICs could continue to cool the reactor for 8 h when the reactor stopped. The cooling period could be extended by replenishing water in the ICs.

Seventeen minutes into the accident, operators of Unit 1 stopped IC-cooling because the cooling rate was too high. The instructions were that the cooling rate of the reactor should not exceed 55 K h^{-1}. About 27 min into the accident, when the temperature and pressure began to return to normal, operators switched one IC on again. By switching the active IC on and off by using valve MO-3A, they controlled the reactor pressure manually.

Fifty-three minutes into the accident, Unit 1 was hit by the second wave of a tsunami, causing the loss of electric emergency power. Out of 13, 12 diesel generators stopped supplying electric power to the power plant. As a result, Unit 1 could no longer receive electric power. Just before that happened, operators had closed valve MO-3A. Because electric power had been lost, it was not possible to open that valve again. Thus, the ICs had become inoperable. When the IC cooling stopped, the reactor's remaining heat flow had decreased to 2% of the heat flow before the accident.

The reactor pressure rose after the tsunami had hit the plant because IC-cooling was no longer active. When the pressure rose approximately 10%, a relief valve atop the reactor opened automatically and steam was vented into the PCV. The relief valve kept the reactor pressure at approximately 70 bar, the saturated vapor pressure of water having a temperature of 284.5 °C. Thus, as long as there was water in the reactor, the reactor temperature could not exceed this temperature. Steam was condensed in the SC and the temperature of water therein rose (see Figure 10.4).

Reactor cooling was maintained by automatic opening and closing of the relief valve to release generated steam. It has been recorded that the radiation level in front of the double door of the reactor building was high at approximately 23.00 h on March 11 (8 h and 14 min into the accident). That is evidence that a substantial amount of radioactive material had leaked into the PCV. It has been estimated by TEPCO that by midnight the reactor's core had lost all water and that a chemical reaction between zirconium and water had occurred. Zirconium was present in the fuel tubes made of Zircaloy. The chemical reaction generated hydrogen and heat.

It has also been recorded that the pressure inside the PCV was 6 bara at approximately 23.50 h (9 h and 4 min into the accident). That means that an opening had been formed in the reactor and that steam had leaked into the PCV. The temperature of the water in the SC had risen to approximately 160 °C.

Molten core material fell out of the reactor on the floor of the PCV between 0.00 and 04.00 h on March 12. It can be assumed that the temperature of the molten core material was 2000 °C. The molten material sank about 65 cm into the reinforced concrete. It cooled down thereby. The thickness of the reinforced concrete was 2.6 m [11].

Pumping of seawater into the PCV by means of a fire engine started at 04.00 h (13 h and 14 min into the accident). The seawater ran down the reactor's circumference. The pressure inside the PCV was, at the time the seawater injection started, approximately 8 bara. The temperature of the water in the SC had risen to approximately 170 °C. The seawater flow amounted to, at the discharge pressure of approximately 8 bara, 5 t h^{-1}. Both the pressure inside the PCV and the seawater flow stayed approximately constant for a period of 10.5 h.

It can be calculated that the decay heat caused the complete evaporation of the water in the seawater flow. The calculation proceeds as follows. The capacity figures of Unit 1 are 460 MWe and 1500 MWth. All figures that are given in this paragraph are approximate figures. The seawater injection by the fire engine at a rate of 5 t h^{-1} started at 04.00 h on March 12. The decay heat was 0.7% of 1500 MWth at that point in time as more than 13 h had elapsed since the reactor had been stopped. And 0.7% of 1500 MWth is 10.5 MW. On taking 3000 kJ for the heating and evaporation of 1 kg of water in seawater, the latter heat flow is able to heat and evaporate 12.6 t of water in seawater per hour. The seawater injection of 5 t h^{-1} did not even come close to matching the decay heat. That means that the temperature of the core material kept rising. The rapid evaporation of the water in the seawater flow also explains that the pressure in the PCV stayed more or less constant at a level of 7–8 bara. The contact between the molten core material on the floor of the PCV and gaseous water was not intimate. Only a small amount of hydrogen could additionally be formed.

The pressure inside the PCV was relieved to about 5 bara by venting to the atmosphere at 14.30 h on March 12 (23 h and 44 min into the accident). This caused the seawater flow to rise to 30 t h^{-1}.

The decay heat could now no longer evaporate all water in the seawater flow. The intimate contact between molten core material on the floor of the PCV and seawater caused the formation of a substantial amount of hydrogen gas and a pressure rise. The high pressure in the PCV lifted the top of the PCV. Next, the shield plug was lifted and a gaseous mixture flowed from the PCV into the space of the fuel exchange floor. Here, it mixed with air containing oxygen. An explosion occurred at 15.36 h on March 12 (24 h and 50 min into the accident). The explosion blew off the roof and cladding on the top part of the building. The explosion has probably been ignited by a spark caused by the falling back of the shield plug. There was an open connection between the PCV and the atmosphere after the escape of the gaseous mixture from the PCV.

Context and Aftermath

The cores of the reactors of Fukushima Daiichi Units 1–3 largely melted in the first three days [11]. The molten material of the reactors of Units 2 and 3 stayed in the reactors. There were hydrogen/oxygen explosions in Units 1, 3, and 4 in the first five days. Unit 4 was not operating at the time of the accident. The explosion in Unit 4 was caused by the leakage of hydrogen gas from Unit 3. Major releases of radionuclides, including long-lived cesium, occurred to air, mainly in mid-March 2011. The main source of radioactive releases was a gas release from Unit 2 on March 15. An explosion did not occur in Unit 2 as it was severely damaged by the explosion in Unit 1 and gases could flow freely to the atmosphere. The electric power supply to Fukushima Daiichi was restored after 10 days. The reactors of Units 1–3 were stable with water addition after 2 weeks and by July 2011 they were being cooled with recycled water from a new treatment plant. Official "cold shutdown condition" was announced in mid-December 2011.

Damage

Three TEPCO employees were killed when the earthquake and the tsunami hit the power station [11]. There have been no direct deaths of radiation sickness from the nuclear accident at Fukushima. Approximately 160 000 people were evacuated from their homes. Only in 2012 limited return was allowed and in October 2013, 81 000 evacuees remained displaced due to government concern about radiological effects from the accident.

The material damage of the accident was substantial.

Remarks

The heart of the matter of the events in Unit 1 of the nuclear power plant Fukushima Daiichi is that, like at Three Mile Island (see Section 10.3), it has not been possible to transfer the remaining heat of the nuclear reactor safely.

The protection of the nuclear reactor of Unit 1 relied on active and procedural safety measures (see Chapter 2). Active process protection starts working upon a signal. Procedural safety measures concern action taken by humans.

The starting up of the diesel generators to supply electric power when the earthquake occurred was the first active safety measure. That measure was successful.

The second measure was the closing of the main steam isolation valve when the earthquake occurred. The measure was successful. Steam raised in the reactor could no longer flow to the turbines.

The introduction of the nuclear control rods into the reactor to stop nuclear fission was the third measure. It was successful as well.

The fourth protection measure was the starting up of two ICs to cool the reactor. That safeguarding method was activated by a signal indicating too high a pressure in the reactor. Although the safety measure was successfully started up, its effect was mitigated by the operators of Unit 1 and the loss of electric power.

The fifth safety measure was procedural and comprised the pumping of seawater into the PCV. The measure was taken late and the seawater flow was small. The temperature of the molten core material kept rising.

The sixth protection measure was also procedural and concerned the venting of the PCV. The measure had an adverse side-effect because the increased seawater flow caused the generation of a substantial amount of hydrogen gas and an explosion.

The primary containment was a passive safety feature. However, it failed due to the high pressure built up by the formation of hydrogen after the venting of the PCV.

10.5 HIGH-TEMPERATURE GAS-COOLED REACTORS (HTGRs)

10.5.1 Introduction

High-temperature gas-cooled reactors (HTGRs) are discussed in this section because they have better safety characteristics than LWRs and relatively large power stations equipped with HTGRs have been built and operated.. See Figure 10.6 in which two reactor types are depicted.

The functioning of these two reactor types will be described briefly. The reactor at the left in Figure 10.6 contains fuel in prismatically shaped elements. It is called prismatic block reactor. The prismatic blocks are made of graphite and are stacked one on top of the other. Graphite is moderator in this reactor. There are vertical channels in the blocks that are filled with small fuel cylinders, also called compacts. The small cylinders contain fuel kernels containing UO_2, uranium carbide, or a mixture of thorium carbide and uranium carbide. The kernels are coated with pyrolytic carbon so as to retain fission products. The diameter of the coated kernels is, e.g. 0.95 mm. Helium gas under pressure passes through channels in the fuel elements and is heated. Hot helium gas can then transfer its heat indirectly in a heat exchanger to generate steam. Helium gas is subsequently recycled to the reactor. Generated steam drives a turbine and the turbine drives an electricity generator.

The reactor at the right in Figure 10.6 is known as pebble bed reactor (PBR). The reactor is filled with pebbles having a diameter of, e.g. 60 mm. Like the small cylinders of the prismatic block reactor, the pebbles contain coated fuel kernels. The fuel kernels can have the same size and composition as described for the prismatic block reactor. A 5-mm outer shell of the pebbles consists of graphite. Graphite is also moderator in this reactor. Helium gas under pressure passes through the bed of pebbles and is heated.

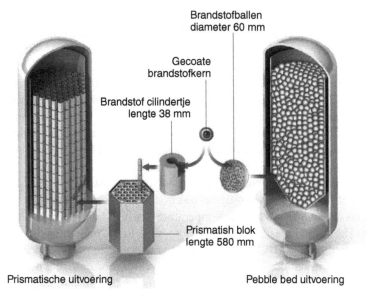

Figure 10.6   Prismatic block reactor at the left and pebble bed reactor at the right. Text of the figure from top to bottom: Brandstofballen, fuel spheres; diameter 60 mm, 60 mm diameter; Gecoate brandstofkern, coated fuel particle; Brandstof cilindertje lengte 38 mm, fuel compact 38 mm tall; Prismatisch blok lengte 580 mm, prismatic block 580 mm tall; Prismatische uitvoering, prismatic configuration; Pebble bed uitvoering, pebble-bed configuration. *Source:* Courtesy of GVO drukkers & vormgevers B.V., Ede, The Netherlands

Hot helium gas can be used in the same way as described for the prismatic block reactor. Pebbles are continuously fed at the reactor top and extracted continuously at the reactor bottom.

The moderator function and the heat carrier function are separated in these two reactor types. Remember that water is both moderator and heat carrier in an LWR. A further aspect is that helium does not interact with neutrons.

Typically, in a PWR, the pressure in the primary circuit is 155 bar and water leaves the reactor at 325 °C. In a PWR, the pressure in the primary circuit must always exceed the saturated vapor pressure at the temperature at which water leaves the reactor. Thus, in order to avoid extremely high pressures, it is not feasible to raise the temperature of the water in the primary circuit of a PWR substantially. In the primary circuits of HTGRs, temperatures higher than 325 °C, e.g. 800 °C, are possible.

The temperature and the pressure in the primary circuit of an HTGR can be chosen independently from each other. That is not possible in an LWR.

However, helium pressure in the primary circuit of an HTGR will usually not be chosen very low to obtain the right conditions for relatively good heat transport and transfer characteristics.

Both reactor types are often claimed to be inherently safe. The development of HTGRs started in the 1950s and 1960s. The nuclear accidents at Three Mile Island (1979), Chernobyl (1986), and Fukushima (2011) had not yet happened then. The main emphasis was, at that time, on the possibility to reach high temperatures in the primary circuit. However, it was also realized that the safety properties of HTGRs were relatively good.

Back in the 1950s and 1960s, there were three reasons to consider the design of HTGRs. The first reason was that the efficiency of nuclear power stations can increase when the temperature at which helium leaves the core rises. The second reason was that it is possible to use helium having a temperature of, e.g. 800 °C, to heat certain reactors in the chemical industry indirectly without conversion to electrical power. For instance, it can heat chemical reactors in which methanol is synthesized from carbon monoxide and hydrogen. The third reason was that HTGRs have better safety characteristics than LWRs.

Concerning the temperatures at which helium leaves and enters the core, two aspects play a role. The first aspect is the Wigner effect [12]. That effect describes the phenomenon that energy is stored in irradiated graphite and can be released suddenly when temperature rises. Such a release of energy leads to a sudden additional temperature rise. The Wigner effect is important at relatively low temperatures and is less important at relatively high temperatures. To avoid the effect, the temperature at which helium enters the core has to be at least approximately 200 °C and preferably at least 300 °C.

The second aspect concerns the investment. To keep the sizes of the steam generators reasonable, the temperature difference across the metal tube walls of the steam generators between helium and water should be substantial. The relatively great difference compensates for the poor coefficient for the heat transfer from helium to the metal wall of the steam generator.

The combination of the two effects leads to, e.g. a temperature of 700 °C at which helium enters the steam generator. Helium would leave the steam generator at, e.g. 300 °C and the temperature difference between the two helium flows would in that case be 400 K.

A further aspect is that fuel is not, like in LWRs, contained in metal tubes. Helium coolant comes into contact with graphite only. A chemical reaction between helium and graphite cannot occur as helium is inert. In

LWRs, water can react with zirconium of the fuel tubes at elevated temperatures and hydrogen gas can be formed.

When graphite at, e.g. 700 °C, comes into contact with air, it will be oxidized. Concerning process safety, this is an aspect to be considered. See the paragraph on PBR process safety in Section 10.5.3.

10.5.2 Safety Aspects of HTGRs

One aspect strikes immediately on comparing the safety aspects of HTGRs and LWRs. A typical power density of a PWR is 230 MW m^{-3} [13]. A typical power density of an HTGR is approximately two orders of magnitude smaller. In other words, HTGRs are, for a given capacity, relatively large and LWRs are relatively small. The core diameter of Peach Bottom Unit No. 1, an HTGR having a capacity of 40 MWe, was 2.79 m (see Table 10.5). A typical PWR having a capacity of 1000 MWe has a diameter of 4.5 m [14]. The difference in power density is caused by the different natures of these two reactor types. The heart of the matter is that light water (normal water) is more effective in capturing neutrons than graphite. The factor between these two moderators is 65 [15]. An LWR cannot function if the fuel rods are placed farther apart from each other, leading to a lower power density. The consequences of the difference in power density will be considered shortly. First, a few words concerning the PWR. If an LOCA (loss of coolant accident) occurs, it can be expected that the shutdown rods are brought in place to stop the heat development due to the fission reactions. However, one still has to deal with decay heat; see Section 10.2.1 for a description of decay heat. That heat can be removed by means of emergency cooling systems. If such systems fail, the situation cannot be kept under control.

Next, a few words concerning an HTGR. If an LOCA occurs, it can be expected that the shutdown rods are brought in place to stop the heat development due to the fission reactions. However, also in this case, one has to still deal with decay heat. The core temperature will in such a case increase to a certain value and then start to fall. Core decay heat can be carried away by conduction, natural convection, and radiation. The criterion is to keep the temperature of the fuel below 1600 °C to avoid liberation of fission products. The reasons of this good safety characteristic are, compared to the LWR, the much larger heat capacity of the reactor per MWth and the much larger outer area per MWth. Both a PBR and a prismatic block reactor should not exceed a certain size to maintain this good safety characteristic. Those sizes will be mentioned when the PBR and the prismatic block reactor are discussed.

There is one further major aspect. If, in the case of a HTGR, the shutdown rods cannot be brought in place when an LOCA occurs, the fission heat can also be kept under control. The temperature of the core then rises and the fission reactions practically stop immediately due to the Doppler effect. At relatively low temperatures, neutrons are effective concerning fission. However, they lose effectivity when the temperature rises because more neutrons are then captured by U-238 or by Th-232. Thus, the Doppler effect results in a negative temperature coefficient, i.e. the core's reactivity decreases when the temperature increases.

Summing up, HTGRs can be designed such that the fuel temperature remains below 1600 °C during serious accident conditions, i.e. helium pipe rupture, simultaneous loss of electrical power supply, and simultaneous failure of the emergency shutdown.

10.5.3 PBR

Description The small spheres containing the fuel are enwrapped by layers of ceramic material (see Figure 10.6). They are combined by pressing them into pebbles. The large spheres are, in the manufacturing process, sintered at a high temperature. It is not necessary to shut a reactor down for fuel exchange as the pebbles are replaced regularly. Shutdown rods can be used to stop nuclear fission. However, when shutting down a reactor, one still has to deal with decay heat.

The PBR is a German development. However, Germany stopped the development in 1989. A small reactor was started up in China in 2000 and was still in operation in 2017. It has a capacity of 10 MWth. It was planned to start up a commercial unit in China in 2017. This Chinese nuclear power plant will be equipped with two PBRs each having a capacity of 250 MWth. The plant's total electric capacity is 210 MWe.

German Experience

General A small power station with 46-MWth and 15-MWe capacity has successfully been in operation at Jülich between 1966 and 1989. The station was indicated as AVR ("Arbeitsgemeinschaft Versuchsreaktor", German for "consortium experimental reactor"). The objective was to obtain experience with a PBR and to collect data for scale-up. Data have been scaled-up and a large power station of 750-MWth and 300-MWe capacity and also equipped with a PBR has, less successfully then AVR, been in operation at Hamm-Uentrop between September 1985 and September 1989. The power station was indicated as THTR-300 ("Thorium Hochtemperaturreaktor", German for "thorium high-temperature reactor").

AVR

Description   Helium was recycled in a primary circuit by two blowers. It passed from the bottom of the pebble bed to the top [16]. Hot helium gas transferred heat indirectly in one steam generator built atop the reactor core. The reactor core was surrounded by a graphite reflector having a wall thickness of 50 cm. Graphite acted as moderator and the thick graphite wall served to prevent neutrons liberated at the fission process to escape from the reactor. The reactor power could be changed by changing the speed of the blowers [17]. This was possible because of a negative temperature coefficient at all operating conditions; the reactor power was therefore proportional to the cooling gas mass flow. Having a negative temperature coefficient means that the reactor's activity decreases when the temperature increases and vice versa. A negative temperature coefficient is caused by the Doppler effect (see Section 10.5.2). For normal operation, the reactor was equipped with four shutdown rods being able to move vertically in graphite tubes. There were no additional control rods. The steam drove a turbine and the turbine drove an electricity generator. Table 10.3 contains technical and operational data of the power plant.

Table 10.3   Operational and technical data of AVR and THTR-300.

Aspect	AVR [18]	THTR-300 [22]
Thermal power (MW)	46	750
Electrical power (MW)	15 [19]	300[a]
Power density (MW m^{-3})	2.6	6[b]
Cooling gas inlet temperature (°C)	275	250
Cooling gas outlet temperature (°C)	950	750
Cooling gas pressure (bar)	10.8	40
Cooling gas mass flow (kg s^{-1})	13	—
Core diameter (m)	3.0	5.6
Average core height (m)	2.8	5.1
Steam pressure (bar)	72	190
Steam temperature (°C)	505	545
Number of shutdown rods	4	42 [23]

— means data not found.
[a] Approximately.
[b] Average.

Fuel Elements   The core consisted of 100 000 spherical graphite fuel elements that contained the fuel in coated particles [18, 20]. These pebbles had a diameter of 6 cm. The pebbles normally contained 1 g of U-235. There were only a few exceptions. The percentages enrichment used were 10, 16.7, and 93 (the last value was, at that time, still allowed). In addition, the fuel elements contained 0, 5, or 10 g of thorium. U-238 or Th-232 can take care of the Doppler effect.

Safety Characteristics [21]   An experiment illustrating safety aspects of the PBR will be described. The purpose of the test was the simulation of the conditions of an LOCA as realistically as possible. Decay heat was simulated by fission heat at the experiment. It is assumed that, at this specific test, the fuel elements contained 1 g of U-235 and no thorium. Some weeks before the start of the experiment, the feeding and extraction of pebbles to, respectively from, the reactor were stopped. The reactivity subsequently decreased and the position of the shutdown rods was adjusted to obtain a reactor capacity of 4 MWth, about 9% of the full-load value. The cooling gas pressure was lowered to 1 bar. The speed of the blowers was adjusted in such a way that the cooling gas mass flow, like the thermal power, was about 9% of the full-load value. After a few days of 4-MWth operation, the reactor temperature distribution basically corresponded to the reactor temperature distribution at full-load operation. The experimental cooling gas outlet temperature was, at that time, approximately 800 °C. Next, an accident (LOCA) was simulated by shutting down the blowers. Reactor temperatures at various locations were recorded for a period of 120 h. All measurements were lower than 900 °C, and, after 100 h, all reactor temperatures decreased as a function of time.

The experiment proved that, in the case of an LOCA and activation of the shutdown rods, decay heat could be removed from the core without forced cooling while unacceptably high core temperatures did not occur. Most of the core heat was transferred to the steam generator by convection and radiation in the period of 120 h. Calculations showed that the result of this experiment could be scaled-up to THTR-300 [16].

The experiment proved that the reactor could deal with the decay heat after the shutdown rods would have been introduced successfully into the core. Introducing the shutdown rods into the reactor running at a capacity of 46 MWth would lower the capacity instantaneously to approximately 4 MWth. Thus, the safeguarding of the reactor by means of the shutdown rods would be active. However, further active process protection, such as emergency cooling systems, would not be required. In the case of LWRs, introduction of control rods would not be sufficient and further active process safeguarding would be necessary. The experiment proved therefore that, in comparison to LWRs, the PBR is a reactor with improved process safety characteristics.

The reason to organize the experiment in the described manner was that it was not practical to remove the helium gas instantaneously from the primary circuit. It would have cost about 3 days to process the helium gas through the gas purification system.

Explanation of the Good Process Safety Properties   The power density of AVR is $2.6\,MW\,m^{-3}$ (see Table 10.3). A typical power density of a PWR is $230\,MW\,m^{-3}$ [13]. Thus, the power density of a typical PWR is almost a factor 90 greater than the power density of AVR. That difference is an important aspect affecting the process protection. AVR has relatively more mass than a PWR to absorb heat and has relatively a larger outer area than a PBR to transfer heat to the surroundings.

A further aspect is the existence of the Doppler effect (see Section 10.5.2). AVR's moderator is graphite.

THTR-300

Description   A prestressed concrete reactor vessel contained both the core and six steam generators [16, 22]. It had an outer diameter of 16 m, a height of 18 m, and a 5-m wall thickness. Uranium dioxide (UO_2) or a combination of UO_2 and thorium dioxide (ThO_2) was used as a fuel. Helium was used as coolant. It was recycled in a primary circuit by blowers. It entered the core at the top and left it at the bottom. Hot helium gas transferred heat indirectly in the steam generators. A total of 42 shutdown rods could be inserted into the pebble bed [23]. The shutdown rods then came into direct physical contact with pebbles. Steam drove a turbine and the turbine drove an electricity generator. Table 10.3 contains technical and operational data of the power plant.

Operational Experience   The power plant has been in operation between September 1985 and September 1989 [24]. Good safety characteristics at regular process conditions have been proven. The time availability was 61% between June 1, 1987, and January 1, 1988, and 51% in 1988. These relatively low figures have been caused by the fuel spheres handling system. Other parts of the plant functioned satisfactorily. The station owners decided to wind-up the activities for financial reasons in 1989.

A Nuclear Incident   An incident occurred with the THTR-300 at Hamm-Uentrop on May 4, 1986 [25]. A pebble that had to be added to the bed got stuck in a pipe. While making attempts to remedy this, an operator erroneously released radioactive gas to the atmosphere. The amount of radioactivity released has been estimated at approximately $90 \cdot 10^6$ Becquerel. On the basis of this amount, the incident could be classified as a minor accident. A Commission of Inquiry reported that 75% of the radioactivity in the vicinity of the power station was caused by the release on May 4, 1986, and 25% was caused by the nuclear accident at Chernobyl on April 26, 1986.

Remarks Concerning Scaling-up Scaling-up the fuel spheres handling system has not been successful. The AVR system to handle fuel spheres was already relatively complicated [26]. However, it functioned satisfactorily. The THTR-300 system to handle fuel spheres has caused a relatively low time availability. The scaling-up factor was 20. Generally, the flow of liquids and gases is easier to control than the flow of particulate materials. Pebbles are a particulate material. Also generally, concerning the design of fuel sphere handling systems, the pressure in the primary circuit is an important aspect. The pressure in the primary circuit of AVR was 10.8 bar. The pressure in the primary circuit of THTR-300 was 40 bar. A fuel sphere handling system implies that fuel spheres are continuously fed from a space having atmospheric pressure into a space having a relatively high pressure. At the same time, hot helium gas is not allowed to flow in the reverse direction. That requires complicated mechanical provisions. A similar remark can be made for the extraction of pebbles.

The direction of the helium flow for AVR was from the bottom to the top of the reactor, whereas it was from the top to the bottom for THTR-300. The reason for this change was that the upper fuel spheres, when high gas velocities were selected, "danced" in AVR. The fuel spheres were stationary at all gas velocities in THTR-300 [16].

Chinese Experience and Plans

General A small power station of 10-MWth and 2.5-MWe capacity is in operation at the Tsinghua University at Beijing Shi since 2000. The station is indicated as HTR-10 (acronym for high-temperature reactor having a thermal capacity of 10 MW). The objective is to obtain experience with a PBR and to collect data for scale-up. Data have been scaled-up and a large power station of 210 MWe is under construction. German experience has also been used. The plant contains two PBRs, each having a capacity of 250 MWth. The power station is indicated as HTR-PM (the acronym PM stands for PBR and modular).

HTR-10 Helium is recycled in a primary circuit. Hot helium gas transfers heat indirectly in one steam generator. The steam drives a turbine and the turbine drives an electricity generator. Table 10.4 contains technical and operational data of the power plant. The fission reaction is moderated by means of graphite.

Safety Characteristics Safety demonstration tests have been carried out. These tests comprised stopping the cooling of the reactor while the reactor produced approximately 3 MWth. That is, the reactor produced at about 30% of the nominal capacity. Stopping the cooling was not followed by the introduction of shutdown rods. The reason to follow this procedure is that the reactor will in that case not be damaged. It would have been possible to stop the cooling while the reactor produced at 100% of the nominal capacity. Likewise, such stopping would then not have to be followed by the introduction of shutdown rods. It is predicted that, in that case, there would be no environmental consequences; however, it is probable that the reactor would be damaged.

HTR-PM Table 10.4 contains technical and operational data of the power plant under construction.

PBR Process Safety The maximum size of a PBR having passive safety characteristics is approximately 250 MWth [29]. That means that the fuel temperature remains below 1600 °C during serious accident conditions,

Table 10.4 **Operational and technical data of HTR-10 and HTR-PM [27, 28].**

Aspect	HTR-10	HTR-PM
Thermal power (MW)	10	2×250
Electrical power (MW)	2.5	210
Power density (MW m^{-3})	2.0	3.2
Cooling gas inlet temperature (°C)	250	250
Cooling gas outlet temperature (°C)	700	750
Cooling gas pressure (bar)	30	70
Cooling gas mass flow (kg s^{-1})	4.3	—
Core diameter (m)	1.80	3
Average core height (m)	1.97	11
Steam pressure (bar)	35	132.5
Steam temperature (°C)	435	567
Number of shutdown rods	—	—

— means data not found.

i.e. helium pipe rupture, simultaneous loss of electrical power supply, and simultaneous failure of the emergency shutdown. Emission of radioactive materials cannot occur. The fission reaction is then practically halted due to a temperature rise and the associated Doppler effect. The decay heat can be transferred to the surroundings. This statement is based on experiments and calculations.

Helium pipe rupture implies air ingress. The oxygen in air will oxidize graphite of the core and the fuel elements in that case. However, tests showed that this oxidation is a relatively slow process because of lack of oxygen supply.

10.5.4 Prismatic Block Reactor

Description The prismatic block reactor is an American development. However, that development was stopped in 1989. At present, there is a test reactor in Japan (see Figure 10.6). The fuel kernels that are used for the PBR can also be used for this reactor type. The kernels are combined by pressing them into hollow cylinders having a height of, e.g. 39 mm. The outer and inner diameters of the small cylinders, also called compacts, are, e.g. 26 and 10 mm. The height of a prismatic block is, e.g. 580 mm. The blocks have a hexagonal cross section. They are stacked one on top of the other to fill the reactor core. The core can have an annular form. The core contains channels for rods to control the nuclear fission. The core also contains channels for the passage of helium gas. The reactor has to be shut down for refueling.

American Experience

General A relatively small power station of 115-MWth and 40-MWe capacity and equipped with a prismatic block reactor has successfully been in operation at Peach Bottom Atomic Power Station at Peach Bottom Township, PA, between 1967 and 1974. The station was indicated as Peach Bottom Unit No. 1. The objective was to obtain operational experience with this reactor type and to collect data for scale-up. The data have been scaled-up and a large power station of 330 MWe capacity and also equipped with a prismatic block reactor has, less successfully than the small power station, been in operation at Fort St. Vrain between 1976 and 1989.

Peach Bottom Unit No. 1 The fuel elements contained coated uranium carbide and thorium carbide particles [30]. The initial fuel loading of the reactor core comprised 220 kg of U-235, 16 kg of U-238, and 1450 kg of

Th-232. The degree of uranium enrichment was thus 93%. That degree of enrichment is no longer allowed today. Helium was recycled in a primary circuit by blowers. It passed through the core consisting of prismatic blocks. Hot helium gas transferred heat indirectly in one steam generator. The reactor core was surrounded by a graphite reflector having a wall thickness of 2 ft (0.61 m). Like for AVR, graphite acted as moderator and the thick graphite wall served to prevent neutrons liberated in the fission process to escape from the reactor. The reactor power was controlled by means of 36 operating rods. Furthermore, there were 19 shutdown rods with an electric drive and 55 emergency shutdown rods that were fuse-operated, i.e. thermally operated. Raised steam drove a turbine, which in turn drove an electricity generator. Table 10.5 contains technical data of the power plant. The power station was exploited commercially. The decision to wind-up the activities has been taken for financial reasons.

An Important Process Learning   The power station operated with Core 1 fuel blocks between 1967 and 1970. These blocks failed as could be noticed by an increase of the radioactivity in the primary system as a function of time and by visual inspection after shutdown. Core 1 blocks were replaced by Core 2 blocks. The latter blocks did not fail. The improvement was

Table 10.5   Operational and technical data of Peach Bottom Unit No. 1 and Fort St. Vrain.

Aspect	Peach Bottom Unit No. 1 [30]	Fort St. Vrain [31]
Thermal power (MW)	115	—
Electrical power (MW)	40	330
Power density (MW m^{-3})	—	—
Cooling gas inlet temperature (°C)	344	404
Cooling gas outlet temperature (°C)	728	777
Cooling gas pressure (bara)	23.8	47.6
Cooling gas mass flow (kg s^{-1})	—	—
Core diameter (m)	2.79^a	—
Core height (m)	2.28^b	—
Steam pressure (bara)	98.6	163.3
Number of shutdown rods	19	—

— means data not found.
[a]Effective.
[b]Active.

achieved by replacing the coating of the kernels containing the fuel by a better coating.

Fort St. Vrain   The plant featured a uranium–thorium fuel cycle [31]. Helium was recycled in a primary circuit by four blowers. It passed down through the core consisting of prismatic blocks. Hot helium gas transferred heat indirectly in 12 steam generators. The reactor core was surrounded by a graphite reflector having a thick wall. Graphite acted as moderator and the thick graphite wall served to prevent neutrons from the fission process to escape from the core. The reactor was equipped with control rods. Raised steam drove a turbine, which in turn drove an electricity generator. Table 10.5 contains technical data of the power plant. The power station was exploited commercially. The decision to wind-up activities has been taken for financial reasons.

Proven Technology   The time availability of the plant was, during the years in which electricity was produced, unsatisfactory. The main reason was that the design chosen for the four helium blowers had not been proven for this application. The helium blowers were, by means of steam turbine wheels, driven by steam raised by the plant itself. Driving helium blowers by electricity would have been a proven design.

Prismatic Block Reactor Safety   The maximum size of a prismatic block reactor having passive safety characteristics is approximately 625 MWth [29]. That means that the fuel temperature remains below 1600 °C during serious accident conditions, i.e. helium pipe rupture, simultaneous loss of electrical power supply, and simultaneous failure of the emergency shutdown. Emission of radioactive materials cannot occur. The fission reaction is then practically halted due to a temperature rise and the associated Doppler effect. The decay heat can be transferred to the surroundings. This statement is based on experiments and calculations.

Helium pipe rupture implies air ingress and graphite oxidation. See the text concerning the process safety of the PBR in Section 10.5.2.

10.5.5   Comparison Between PBR and Prismatic Block Reactor

Refueling   A prismatic block reactor must be stopped for refueling. The Fort St. Vrain reactor has been in operation for 10 years. Refueling has occurred three times during these 10 years. It has probably been combined with other activities. The first and the third refueling lasted several months

[31]. LWRs also need refueling. Usually, it occurs once per annum and comprises the replacement of one third to one quarter of the fuel. It lasts 2–3 weeks, and maintenance activities are also carried out in this period [14]. Refueling a prismatic block reactor needs more attention than refueling an LWR. The reason is that fuel of an LWR can be handled while the fuel is immersed in water. Water prevents the emission of radioactivity. The fuel of a prismatic block reactor must be handled by robots.

Stopping a PBR for refueling is not necessary as the fuel is replaced continuously.

Fuel Handling System   The fuel is stationary in a prismatic block reactor. It is replaced when refueling occurs. Fuel is handled continuously in a PBR and that is the main cause of the unsatisfactory availability of THTR-300. The selection of proven technology is not possible in this case as the solid handling system is unique for the PBR. The performance of the solid handling system in the Chinese commercial nuclear power plant having two PBRs can probably be checked in the coming years.

Prototype Functioning   Prototypes of both the PBR and the prismatic block reactor have been operated. Neither of these two prototypes has functioned satisfactorily. In the case of the PBR, the relatively poor availability was caused by the reactor itself. In the case of the prismatic block reactor, the relatively poor availability was caused by technical parts not directly related to the reactor. The performance of such technical parts can be improved. Therefore, the prismatic block reactor appears to be a better option than the PBR.

A further reason for this choice is the fact that the maximum size of a prismatic block reactor having passive safety characteristics is approximately 625 MWth, whereas that figure is approximately 250 for a PBR (see also Section 10.6).

10.6   COMPARISON BETWEEN LIGHT WATER REACTORS (LWRs, i.e. PWRs AND BWRs) AND HTGRs

Safetywise, an HTGR can be passively protected, whereas an LWR cannot.

An LWR having a capacity of 2500 MWth is more or less standard. The maximum size of a prismatic block reactor having passive safety properties is about 625 MWth. Thus, to match the capacity of a standard LWR, four prismatic block reactors would be needed in parallel. That could still be a practical option.

A nuclear power station having a prismatic block reactor would probably not need extensive emergency cooling systems. A nuclear power station having an LWR will need extensive emergency cooling systems.

The maximum size of a PBR having passive safety properties is about 250 MWth. Thus, to match the capacity of a standard LWR, 10 PBRs would be needed in parallel. That would be a less practical option.

A last remark: if an LOCA occurs in an HTGR, air comes into contact with graphite of the core. Such a contact results in oxidation. See the paragraph on PBR process safety in Section 10.5.3.

REFERENCES

[1] Bogtstra, F.R. (2013). *Nuclear Power – How About It?* 23–24. Bergen NH, The Netherlands: BetaText (in Dutch).

[2] Bogtstra, F.R. (2013). *Nuclear Power – How About It?* 60. Bergen NH, The Netherlands: BetaText (in Dutch).

[3] Bogtstra, F.R. (2013). *Nuclear Power – How About It?* 47. Bergen NH, The Netherlands: BetaText (in Dutch).

[4] Ishikawa, M. (2015). *A Study of the Fukushima Nuclear Accident Process*, 11. Tokyo, Japan: Springer Japan.

[5] Kemeny, J.G. et al. (1979). *Report of the President's Commission on the Accident at Three Mile Island*, 31. Washington, DC: U.S. Government Printing Office.

[6] Kemeny, J.G. et al. (1979). *Report of the President's Commission on the Accident at Three Mile Island*, 12. Washington, DC: U.S. Government Printing Office.

[7] Kemeny, J.G. et al. (1979). *Report of the President's Commission on the Accident at Three Mile Island*, 81–149. Washington, DC: U.S. Government Printing Office.

[8] Kemeny, J.G. et al. (1979). *Report of the President's Commission on the Accident at Three Mile Island*, 107. Washington, DC: U.S. Government Printing Office.

[9] Ishikawa, M. (2015). *A Study of the Fukushima Nuclear Accident Process*, 101. Tokyo, Japan: Springer Japan.

[10] Ishikawa, M. (2015). *A Study of the Fukushima Nuclear Accident Process*, 45–51, 91–108. Tokyo, Japan: Springer Japan.

[11] World Nuclear Association (2016). Fukushima Accident, Internet.

[12] Nuclear Energy Directorate (Commisariat à l'énergie atomique) (2006). *Gas-Cooled Nuclear Reactors*, 29. Gif-sur-Yvette, France: CEA.

[13] Association of German Engineers (VDI) – The Society for Energy Technologies (Publ.) (1990). *AVR – Experimental High-Temperature Reactor*, 4. Düsseldorf, Germany: VDI-Verlag GmbH.

[14] Bogtstra, F.R. (2013). *Nuclear Power – How About It?* 58. Bergen NH, The Netherlands: BetaText (in Dutch).

[15] Bogtstra, F.R. (2013). *Nuclear Power – How About It?* 48. Bergen NH, The Netherlands: BetaText (in Dutch).

[16] Cleve, U. (2009). The technology of high-temperature reactors – design – building – start-up – operation of AVR (Jülich) and THTR-300. ATW – *International Journal for Nuclear Power* 54: 776–785. (in German).

[17] Association of German Engineers (VDI) – The Society for Energy Technologies (Publ.) (1990). *AVR – Experimental High-Temperature Reactor*, 240. Düsseldorf, Germany: VDI-Verlag GmbH.

[18] Association of German Engineers (VDI) – The Society for Energy Technologies (Publ.) (1990). *AVR – Experimental High-Temperature Reactor*, 89. Düsseldorf, Germany: VDI-Verlag GmbH.

[19] Association of German Engineers (VDI) – The Society for Energy Technologies (Publ.) (1990). *AVR – Experimental High-Temperature Reactor*, 247. Düsseldorf, Germany: VDI-Verlag GmbH.

[20] Association of German Engineers (VDI) – The Society for Energy Technologies (Publ.) (1990). *AVR – Experimental High-Temperature Reactor*, 75–76. Düsseldorf, Germany: VDI-Verlag GmbH.

[21] Association of German Engineers (VDI) – The Society for Energy Technologies (Publ.) (1990). *AVR – Experimental High-Temperature Reactor*, 245–258. Düsseldorf, Germany: VDI-Verlag GmbH.

[22] Nickel, H., Hofmann, K., Wachholz, W., and Weisbrodt, I. (1991). The helium-cooled high-temperature reactor in the Federal Republic of Germany: safety features, integrity concept, outlook for design codes and licensing procedures. *Nuclear Engineering and Design* 127: 181–190.

[23] Association of German Engineers (VDI) – The Society for Energy Technologies (Publ.) (1990). *AVR – Experimental High-Temperature Reactor*, 335. Düsseldorf, Germany: VDI-Verlag GmbH.

[24] Association of German Engineers (VDI) – The Society for Energy Technologies (Publ.) (1990). *AVR – Experimental High-Temperature Reactor*, 311–337. Düsseldorf, Germany: VDI-Verlag GmbH.

[25] Anonymous (1986). Sparkling eyes. *Der Spiegel* 24: 28–30. (in German).

[26] Association of German Engineers (VDI) – The Society for Energy Technologies (Publ.) (1990). *AVR – Experimental High-Temperature Reactor*, 187–202. Düsseldorf, Germany: VDI-Verlag GmbH.

[27] Yuliang, S. (2017). HTR-PM Project Status and Test Program, Internet.

[28] Li, F. (2017). HTR Progress in China, Internet.

[29] AREVA Inc. (2014). *AREVA HTGR*, 6. Charlotte, NC, USA: AREVA Inc.

[30] Everett, J. III and Kohler, E.J. (1978). Peach Bottom Unit No. 1: a high performance helium-cooled nuclear power plant. *Annals of Nuclear Energy* 5: 321–335.

[31] Brey, H.L. (1991). Fort St. Vrain operations and future. *Energy* 16: 47–58.

INDEX